Advance Praise for
STEM21

"Despite over 40 years of concerted efforts to broaden participation in the fields of Science, Technology, Engineering, and Mathematics (STEM), the chronic underrepresentation of certain groups in these fields is leaving gaping shortages in the growing demand for technical talent. This edited volume demonstrates how to catalyze a transgressive educational reform revolution that ruptures prevailing norms and leverages the strengths of all stakeholders including those who would benefit most from the reforms.

The National Society of Black Engineers, one of the largest student-governed associations based in the United States, wholeheartedly endorses this work as it aligns directly with our mission and strategic plan to triple the number of Black engineers the nation's colleges and universities graduate annually by 2025. A STEM-prepared, confident, and diverse high school graduate is on a glide path to become a successful, secure, and culturally-responsive engineering graduate who is then equipped to solve complex challenges. This volume brilliantly provides the key frameworks, strategies, and best practices to achieve these ends at scale."

—*Karl W. Reid, Executive Director, National Society of Black Engineers*

"From curriculum to community, this volume illustrates the power of the STEM approach. Far more than an integration of subjects, STEM implies empowerment. Problem-solving in classroom contexts nurtures innovators in society. The diverse contributions here can inform every stakeholder who cares about equity and social justice."

—*Juliana Texley, NSTA Past President 2015*

"This volume presents a series of responsive and innovative approaches that expand minoritized students' opportunities to learn STEM. Drawing on bell hooks' vision of transgressive practice, the volume includes chapters that blend theory and practice across a variety of educational contexts, and the editors synthesize them to put forth a framework for Transgressive STEM Teaching. What is also particularly notable is the way the book integrates deeply personal stories throughout. As such, the text itself embodies a pedagogy of vulnerability, a transgressive STEM practice and marginalized discourse in STEM and STEM education. I highly recommend this book for teachers, administrators, researchers, and community partners."

—*Sara Tolbert, Associate Professor of Science Education, University of Arizona*

STEM21

sj Miller & Leslie David Burns
GENERAL EDITORS

Vol. 10

The Social Justice Across Contexts in Education series is
part of the Peter Lang Education list.
Every volume is peer reviewed and meets
the highest quality standards for content and production.

PETER LANG
New York • Bern • Berlin
Brussels • Vienna • Oxford • Warsaw

STEM21

Equity in Teaching and Learning to Meet Global Challenges of Standards, Engagement and Transformation

Edited by
Joy Barnes-Johnson and Janelle M. Johnson

PETER LANG
New York • Bern • Berlin
Brussels • Vienna • Oxford • Warsaw

Library of Congress Cataloging-in-Publication Data

Names: Barnes-Johnson, Joy, editor. | Johnson, Janelle M., editor.
Title: STEM21: equity in teaching and learning to meet global challenges of
standards, engagement and transformation / edited by Joy Barnes-Johnson
and Janelle M. Johnson.
Description: New York: Peter Lang, 2018.
Series: Social justice across contexts in education; vol. 10
ISSN 2372-6849 (print) | ISSN 2372-6857 (online)
Includes bibliographical references and index.
Identifiers: LCCN 2018013308 | ISBN 978-1-4331-5137-8 (hardback: alk. paper)
ISBN 978-1-4331-5138-5 (paperback: alk. paper)
ISBN 978-1-4331-5139-2 (ebook pdf)
ISBN 978-1-4331-5140-8 (epub) | ISBN 978-1-4331-5141-5 (mobi)
Subjects: LCSH: Science—Study and teaching—Social aspects—United States. |
Educational equalization—United States. | Education, Urban—
United States. | Social justice—United States.
Classification: LCC LB1585.3.S74 2018 | DDC 507.1—dc23
LC record available at https://lccn.loc.gov/2018013308
DOI 10.3726/b13840

Bibliographic information published by **Die Deutsche Nationalbibliothek**.
Die Deutsche Nationalbibliothek lists this publication in the "Deutsche
Nationalbibliografie"; detailed bibliographic data are available
on the Internet at http://dnb.d-nb.de/.

The paper in this book meets the guidelines for permanence and durability
of the Committee on Production Guidelines for Book Longevity
of the Council of Library Resources.

© 2018 Peter Lang Publishing, Inc., New York
29 Broadway, 18th floor, New York, NY 10006
www.peterlang.com

All rights reserved.
Reprint or reproduction, even partially, in all forms such as microfilm,
xerography, microfiche, microcard, and offset strictly prohibited.

Printed in the United States of America

To the ones who teach us and to the ones we teach ...

Table of Contents

List of Illustrations	ix
List of Tables	xi
List of Contributors	xiii
Acknowledgements	xv
List of Abbreviations	xvii

Introduction   1
    Joy Barnes-Johnson and Janelle M. Johnson

Section One: Standards and Pedagogy: Applying Research to Practice   **17**
Chapter One: 21st c LEASE: Language of Equity and Access to STEM Education   19
    Joy Barnes-Johnson
Chapter Two: Hip-Hop Pedagogy as a Framework to Support the
    Development of Science Geniuses   55
    Edmund S. Adjapong
Chapter Three: Seeding the Future: Social-Justice Driven STEM Education   77
    Christian Konadu Asante, Jacqueline DeLisi,
    Megan McKinley, and Michael Barnett

Section Two: Engagement: Extended Learning Opportunities   **93**

Chapter Four: New Roles and Relationships in Urban STEM Learning Environments: How the Peer-Enabled Restructured Classroom Enhances Equity and Access   95
Leslie S. Keiler and Kathleen Robbins

Chapter Five: Early Engagement in Research as a Tool for Broadening Science Participation   114
Cassie Xu, Robert Newton, Margaret Turrin, and Susan Vincent

Chapter Six: Promoting Middle School Students' Motivation and Persistence in an After-School Engineering Program   138
Srinjita Bhaduri, Alexandra Gendreau, Varsha Srikanth Koushik, Tammy Sumner, John Ristvey, and Randy Russell

Section Three: Transformation: Transgressive Practices along the Journey   **163**

Chapter Seven: Engaged Interdisciplinary Literacy: Research and Practices of Secondary STREAM   165
Joy Barnes-Johnson

Chapter Eight: Transformative Education Pathways to Improve Health Literacy, STEM Learning, and Youth Outcomes   188
Gretchen E. L. Suess, Joanna Chae, and Sharon Lewis

Chapter Nine: Institutional Capacity Building for STEM Teacher Education at an Urban Commuter University   205
Janelle M. Johnson, Roland Schendel, Elizabeth McClellan Ribble, and Hsiu-Ping Liu

Implications and Conclusions   225
Joy Barnes-Johnson and Janelle M. Johnson
Appendices   249
Index   261
Conceptual Framework Theorists Index   267

Illustrations

Figure 1.1:	Nested history of the development of equitable STEM education	40
Figure 2.1:	Border crossing pedagogy	57
Figure 6.1:	Notebook	147
Figure 6.2:	Disasterville	149
Figure 7.1:	Teaching slides: Science & technology	180
Figure 7.2:	Teaching slides: Engineering & mathematics	181
Figure 7.3:	Teaching slides: 21st century society	182
Figure 9.1:	Population vs. Survey Respondents	211
Figure 9.2:	Role of Finances	213
Figure 10.1:	Network Theory of Successful STEM Student Identity Development	241

Tables

Table 1.1:	21st century education legislation equity links to STEM competitiveness initiatives	27
Table 1.2:	Banksian multicultural education: Short research timeline	30
Table 1.3:	Applying Tanner's equitable teaching: Responses to strategic self-assessment questions	36
Table 1.4:	Highlighted chapter vocabulary	45
Table 6.1:	Key features of the cohorts and settings for the two UAV curriculum deployments	146
Table 6.2:	Planning a rescue storyboard template	152
Table 6.3:	Research constructs, their definitions, and example prompts used in the flight logs	153
Table 7.1:	STEM-21 unit curriculum design	172
Table 7.2:	STEM-21 teaching and learning goals	174
Table 7.3:	PEAS & Qs portfolio rubric	176
Table 7.4:	Lesson plan BIG IDEAS (month 1 view)	178
Table 10.1:	Descriptions of STEM equity indicators (by subtext)	237
Table 10.2:	Equity—The 4th dimension of the 3D framework of STEM education, Teacher focus	239
Table 10.3:	Equity—The 4th dimension of the 3D framework of STEM Education, Student focus	240

Table Appendix I.1: PERC class target behaviors	250
Table Appendix II.1: C.L.A.S.S. indicators	252
Table Appendix III.1: COMPASS framework for adult stakeholder evaluation	254
Table Appendix IV.1: STEM engagement rubric	256

Contributors

1. Adjapong, Edmund S.—*Seton Hall University*
2. Asante, Christian Konadu—*Boston College*
3. Barnes-Johnson, Joy—*Princeton (NJ) Public Schools*
4. Barnett, Michael—*Boston College*
5. Bhaduri, Srinjita—*University of Colorado Boulder*
6. Chae, Joanna—*Netter Center for Community Partnerships, University of Pennsylvania*
7. DeLisi, Jacqueline—*Education Development Center*
8. Gendreau, Alexandra—*University of Colorado Boulder*
9. Johnson, Janelle M.—*Metropolitan State University of Denver*
10. Keiler, Leslie S.—*York College, The City University of New York*
11. Koushik, Varsha Srikanth—*University of Colorado Boulder*
12. Lewis, Sharon—*Perelman School of Medicine, University of Pennsylvania*
13. Liu, Hsiu-Ping—*Metropolitan State University of Denver*
14. McClellan Ribble, Elizabeth—*Metropolitan State University of Denver*
15. McKinley, Megan—*Boston College*

16. Newton, Robert—*Lamont-Doherty Earth Observatory/Earth Institute, Columbia University*
17. Ristvey, John—*University Corporation for Atmospheric Research (UCAR)*
18. Robbins, Kathleen—*Bronx Early College Academy*
19. Russell, Randy—*University Corporation for Atmospheric Research (UCAR)*
20. Schendel, Roland—*Metropolitan State University of Denver*
21. Suess, Gretchen E. L.—*Netter Center for Community Partnerships & Department of Anthropology, University of Pennsylvania*
22. Sumner, Tammy—*University of Colorado Boulder*
23. Turrin, Margaret—*Lamont-Doherty Earth Observatory/Earth Institute, Columbia University*
24. Vincent, Susan—*Young Women's Leadership School of East Harlem, Retired*
25. Xu, Cassie—*Lamont-Doherty Earth Observatory/Earth Institute, Columbia University*

Acknowledgements

This collection is the result of our wondering about why equity presents so many challenges in the STEM education community. Normalizing questions is a vitally important practice in the science world. We wanted to bring audiences to the edge of this question. We hope we succeeded in drawing out even more questions for both practitioners and researchers.

We tried to showcase theory into practice with each chapter arriving at the end, which "flips" these notions.

With sincere gratitude we extend thanks to the Peter Lang family for encouraging us to press through and bring this work to the forefront. We believe in being stretched for the sake of our students, good teaching and strong(er) communities. To our contributing authors, we sincerely appreciate the time you have invested to share your works with us as editors. We are honored to share this work with its readers.

Abbreviations (by chapter, in order of use)

Introduction

Science Technology Engineering and Mathematics, STEM
American Association for the Advancement of Science, AAAS
English as a Second Language, ESL
Opportunities to Learn, OTL
Pre-kindergarten—grade twelve, PK-12

Chapter 1

Language of Equity and Access to STEM Education, LEASE
Technology, Entertainment, Design Talk, TED Talk
Keep Our Promise to America's Children and Teachers Act, KEEP Our PACT ACT
Elementary and Secondary Education Act, ESEA
Individuals with Disability Education Act, IDEA
No Child Left Behind, NCLB

Every Student Succeeds Act, ESSA
America Creating Opportunities to Meaningfully Promote Excellence in Technology, Education, and Science, America COMPETES Act
Common Core of Data, CCD
Equitable Science Teaching and Learning, EST/L
Culturally Relevant/Responsive Pedagogy; Critical Race Pedagogy, CRP
Opportunities to Learn, OTL
American Association for the Advancement of Science, AAAS
Critical Race Theory, CRT
Next Generation Science Standards, NGSS
Early Childhood Longitudinal Study, ECLS-K
Equitable Science Teaching, EST
Partnership in Education and Resilience, PEAR
Dimensions of Success, DOS
Theory into Practice, TIP
Equitable Science Education, ESE
Science Technology Engineering Art and Mathematics, STEAM
Science Technology Reading Engineering Art and Mathematics, STREAM
Limited English Proficient/cy, LEP
Empowering Science Technology Engineering Art & Mathematics, ESTEAM
Equitable Science Technology Righting/Representing, Engineering, Art and Mathematics Teaching, Eq-STREAM Teaching
Deoxyribonucleic acid, DNA
Near Field Communication, NFC
High School Life Science Standard Numbers, HS-LS#-#

Chapter 2

Historically Black College and University, HBCU
Predominantly White Institution, PWI
Master of Ceremony, MC
Disc Jockey, DJ
Science Technology Engineering Arts Mathematics, STEAM
Bring Attention to Transforming Teaching, Learning and Engagement in Science, BATTLES
Master of Content and Classroom, MC^2

Chapter 3

Light-emitting diode, LED
Out of school time, OST
Social Justice Driven Science Technology Engineering Mathematics, STEMJ
Science Teaching Efficacy Beliefs Instrument, STEBI
Science Teaching Outcome Expectancy, STOE
Teacher Efficacy and Attitudes toward STEM, T-STEM
English Language Learners, ELL
Massachusetts Institute of Technology, MIT

Chapter 4

Peer Enabled Restructured Classroom, PERC
Teaching Assistant Scholars, TAS
Chemistry Learning for Academic Success in Science/Core-Chemistry Lab and Academic Skills for Success in Science, CLASS
Problem-based Learning, PBL
Living Environment, LE
English Language Learners, ELL

Chapter 5

Secondary School Field Research Program, SSFRP
Lamont-Doherty Earth Observatory, LDEO
New York State's Department of Environmental Conservation, NYS DEC
Palisades Interstate Park Commission, PIPC
National Oceanic and Atmospheric Administration, NOAA
Hudson River National Estuarine Research Reserve, HR-NERR
New York City, NYC
The Young Women's Leadership School, TYWLS
Frederick Douglass Academy I, FDA-I
National Science Foundation, NSF
Research One, R1
State University of New York, SUNY

City University of New York, CUNY
Principal Investigator, PI
Matrix Laboratory, MATLAB (Proprietary computer software)
Geospatial Positioning System, GPS

Chapter 6

Unmanned Aerial Vehicles, UAV
Innovative Technology Experiences for Students and Teachers, ITEST
National Society of Black Engineers, NSBE
Summer Engineering Experience for Kids, SEEK
United States, US
National Science Foundation, NSF
Out of school time, OST
Accreditations Board for Engineering and Technology, Inc., ABET
Next Generation Science Standards, NGSS
Division of Research on Learning in Formal and Informal Settings, DRL

Chapter 7

Career and Technical Education, CTE
Advanced Placement, AP
Professional Learning Community, PLC
Individualized Education Plan, IEP
Next Generation Science Standards, NGSS
Science Technology Engineering and Mathematics in a 21st Century Society, STEM—21
Essential Questions, EQ
Performance products, effort, achievement, submission standards, and questions, PEAS&Qs
National Research Council, NRC
National Science Teachers Association, NSTA
Science Technology Reading/wRiting/Rhyming/Righting/Research Engineering Art and Mathematics, STR*EAM

Chapter 8

Health Sciences Education Pipeline Program, Pipeline
University of Pennsylvania, Penn
Out of school time, OST
Every Student Succeeds Act, ESSA
Next Generation Science Standards, NGSS
Extended Learning Afterschool and Summer, ELAS
College Access and Career Readiness, CACR

Chapter 9

Metropolitan State University of Denver, MSU Denver
National Science Foundation, NSF
Hispanic Serving Institution, HSI
Colorado Alliance for Minority Participation, CO-AMP
Colorado-Wyoming Alliance for Minority Participation, CO-WY-AMP
Colorado Association of Black Professional Engineers and Scientists, CABPES
Colorado I Have a Dream Foundation, CIHADF
Center for Advanced STEM Education, CASE

Implications and Conclusions

Technology, entertainment and design, TED
Mentorship, M
Social Justice, SJ
Student Agency, SA
Teacher Agency, TA
Student Inquiry, SI
Interdisciplinarity, I
Place-based education, PBE
STEM Ecosystem Health, ECO
Standards-based, SB
Empowering Identity Formation, ID

Introduction

JOY BARNES-JOHNSON[1] AND JANELLE M. JOHNSON[2]

Science, technology, engineering and mathematics (STEM) content is often celebrated as that which will answer the many questions we have as a society—problems we face in economics,[1] in the environment, and in education are purportedly solved using STEM tools. So, as a society we have attempted to make our students better citizens by building programs of concentrated STEM teaching and learning. We continue to attempt to measure student learning outcomes with standardized testing,[2] knowing that these statistics reveal large and troubling gaps. Industry and government alike respond with urgent calls for diversity, but STEM stakeholders are often unsure about the most appropriate actions they can or should take. In spite of the vaulted "objectivity" of science, its habits and traditions are somehow insufficient for addressing equity issues. The former[3] belief that in a democratic society there would be an equalizing force with which sociopolitical

---

1 Joy Barnes-Johnson, Science Educator
  Princeton (NJ) Public Schools
  joybarnesjohnson@princetonk12.org
2 Janelle M. Johnson, Assistant Professor of Secondary Education
  Metropolitan State University of Denver
  jjohn428@msudenver.edu

tools could grant every human access to liberty and navigate pathways to industry has not been enough to sustain the vision of democratic education, to support a diversity of perspectives in STEM or to build traction for STEM-related creative endeavors. This has not proven possible across professional lines nor across academic ones. Instead, science has been used to reinforce stereotypes and justify egregious crimes against humanity throughout history and at present.

Science is not the sole culprit in creating this intellectual climate. Technology and mathematics have also been used to manipulate information and distort understanding. In many ways, technologies themselves are outgrowing their original purpose of making humans more efficient. Various technologies have now taken on a more primary role of entertaining us and/or making it easier for us to entertain our youth rather than helping us solve problems of societal magnitude or significance. In previous generations, practices of outdoor play and discovery took precedence over indoor manipulation of electronics; the research and opinions of experts[4] call for us to revisit these child-rearing norms. In our generation, appetites for flexible spacing and industrial aesthetics have demanded that art and architecture formerly created only for estoteric value or appreciation—an outdated symbol of advanced thinking and modernity—be replaced by that which is engineered to be beautifully awe inspiring, multi-functional, and utilitarian. Today, we use interactive mathematical technologies in the form of fitness trackers and mobile applications as a mechanism to quantify ourselves and share unprecedented amounts of deeply personal information with marketing agencies and commercial entities at no cost to the manufacturers of this technology; we actually buy tools that feed this "machine," giving access to our homes and lives 24–7 through automation and surveillance applications. Have we gone wrong at home and in school? We know that having access to technology and information helps some individuals, but certainly not all, or even most.

The lines are blurred. It is both exciting and terrifying to think about how much data are generated and exploited in our everyday lives. If we consider emergent discourses around algorithm fairness,[5] information use by governments (predictive policing[6] for example), and BIG DATA,[7] we must take pause and challenge—even protest—how information is being collected, analyzed and processed to communicate ideas to the public. As we gain more experience with all of it, we are able to understand that "unintentional bias is encoded via disparate impact, which occurs when a selection process has widely different outcomes for different groups, even as it appears to be neutral."[8] This is the current context for STEM.

Practitioners and theorists alike debate the definition of STEM. Is it the sum of its four component domains (science, technology, engineering, and math) or does it represent something much bigger than that? Could it be the new and innovative construct of education that engages teachers and students in real life, authentic, and interdisciplinary problem solving? There are many of us who, despite finding our footing more strongly in one of the STEM domains, believe wholeheartedly in a larger vision of STEM. We believe for example in STEM education's potential to tell seamless stories about interrogated observations and answered questions; its ability to facilitate learning through inquiry, and its promise to help today's youth develop competencies for solving society's problems, present and future. The professional organizations representing all the STEM domains espouse this belief as well, as do the latest sets of standards. One only has to peruse the websites for organizations like the National Council of Teachers of Math, or the International Society for Technology in Education for evidence of this important shift. The National Research Council's vision of three-dimensional learning forms the bedrock of the Next Generation Science Standards—practices, crosscutting concepts, and disciplinary core ideas. These three dimensions were outlined to help teachers and students gain and revise knowledge. Unfortunately, one also only has to look at any news source or assessment data set to problematize the huge gaps between those who have negotiated and acquired dominant[9] knowledge from those who have not. There are many types of gap that could be discussed around media coverage and attainment (educational and employment) including achievement, opportunity and value[10] to name a few.

The status quo will not take us successfully into the future. Each of the STEM professional organizations has in some way attempted to tackle this huge problem. We can see evidence of it in their position statements, and in their standards' appendices. However, those documents are on the margins—they are supplements. Equity and inclusion, while deemed important, represent an afterthought. Student learning is thought of in terms of "regular students,"[11] and then "those students." We know who *those students are*, and this volume is about them. It is about social justice and shifting the status quo. Each section of the book includes perspectives on these challenges relevant for pre-service teachers, in-service teachers and out-of-school time (informal/community-based) teaching; each section is contextualized by personal stories of the editors. Editorial reflections that precede each chapter preview and synthesize some of the big ideas of STEM equity discourse outlined in the chapter text.

Much of the literature on the underrepresentation of certain groups in STEM and STEM education reflects a deficit view of non-dominant groups.[12] This volume shifts that focus to student strengths with an equity-based lens that critically examines program outcomes rather than inputs, moving past "good intentions" and design. Strengths-based teaching develops pedagogical approaches that take into account the knowledge and abilities of local families to offer more significant learning opportunities for children in schools;[13] inviting teachers to see students as "whole" people, rather than simply as students in classrooms may facilitate educators' engagement of and with students. In this way, it may be possible to develop capacity, more holistically. Increasing the contributions of diverse communities[14] toward whole-school activities, designing instruction relevant to students and communities through place-based education[15] and viewing linguistic and cultural diversity as resources, rather than as problems[16] is a necessary goal for the 21st century; all of these actions can/must be undertaken inside and outside of traditionally conceived school settings. New definitions and uses of school space and time are on the horizon, a notion easily conceived as "scholastic elasticity."[17] The authors in this volume research a broad range of formal and informal contexts, utilize diverse methodologies, and are all connected by a focus on STEM equity making more meaningful connections to school-like learning by providing flexible options for educative purposes—border crossing.

What is Transgressive STEM Teaching?

In her now seminal work *Teaching to Transgress* (1994), bell hooks invites educators all along the continuum of best practices to think about the most meaningful ways to teach. We take up this invitation and extend it to the current generation, asking questions that hope to remind us of the foundation for better pedagogy that was laid many years before now. At its core, transgressive teaching empowers every stakeholder in the learning space to find their own place and voice, to be self-actualized. Drawing from the work of Giroux and McLaren (1993), we are reminded that as critical pedagogues, this work—scrutinizing educative practices in service to multiculturalism—is a rather essential[18] practice that requires us to blend theories in order to move them into our everyday practices.[19] The outcomes of this critical reflection are: new language, ruptured disciplinary boundaries, decentered authority, agency, power, and struggle. These contributions to transformative pedagogy are only possible through collaboration and border crossing; this volume is framed in that tradition. Informed by our daily experiences teaching and grappling with the challenges of learning to teach in transgressive/transformative ways, we present our perspectives on the vision of engaged pedagogy, high standards of equitable schools and sustainable reform.

Who do we expect to benefit from this work? Stakeholders along the cradle to career continuum will benefit from reading and sharing this volume within their professional communities of practice. We intend for this book to be of vital importance in educational leadership. Administrators and higher education professionals may gain insights about observable practices that develop the dispositions, skills and attitudes that embrace equity in STEM fields and beyond. As a promising mechanism for evaluation that has been largely absent from mainstream educational discourse, these practices will likely distinguish exemplary programs and teachers from those less well suited to serve diverse students. Most directly, however, we expect educators working with pre-service teachers and administrators to take away research-based teaching paradigms that are themselves transgressive, supporting intentionally inclusive STEM pedagogy. We believe this represents a much needed fourth dimension in STEM, and in institutions of learning in general—embedding equity and inclusion in the very structures of our work. Afterthoughts and supplementary documents are helpful, but they are not transformative.

Educators working in formal teaching and learning spaces will benefit from examples provided of high leverage practices. Because practitioners have contributed to this volume, examples from K-12, undergraduate and graduate level teaching and learning provide perspective on immediately implementable changes. Recognizing that action research and case-study methods may not be entirely applicable in every setting, we believe these studies provide a pathway for individual and whole program review. We hope that this volume becomes part of the dialogue of *Project 2061's Science for All Americans* (AAAS, 1989/2009), the National Research Council's *Framework for K-12 Science Education* (2012) and the Next Generation Science Standards (2013). We are embracing a new language and blurring lines between the empirical and the theoretical ... but isn't that STEM?

Our Stories: Editors' Positionality Statements

Positionality statements are critical components of ethical research.[20] Positionality statements are themselves intentional disclosures about the inputs that lead to written or reported outcomes. They are reflexive and, in many ways, transformative. Understanding that by declaring and owning our life experiences as significant forces in our own development, we provide short biographical insights about our becoming equity educators and stakeholders. This volume is laid out to include our, the editors', stories as tangentially related touch points to the STEM education research being shared by each contributor. We foreground each chapter with portions of our individual stories, as we believe these stories are reproduced in the "telling" of STEM teaching and learning research. The short editorial vignettes inform readers about our choice as editors to include each contributing author's article/research report. Italics are used to distinguish our thoughts from those of the contributing others. From these opening disclosures, we are attempting to relate who we are individually to the work of others within and outside of our various communities of practice as rational choices.[21] Our inclusion of these narratives before the research reports of the contributing authors is designed to advance the discourse. As editors, we have made the decision to replace colonizing language in contributors' work to reflect transgressive language;[22] we realize that a lag often exists between the adoption of language and the observation of practices that are truly transformative. Further lags exist between quality execution of transgressive practice and the open publication of resources useful for those trying to use transformative language and practices in their working environments. As editors, we fully understand the complexity of this work—work we acknowledge is typically reflective of sometimes oppressive classifications and hierarchies that exist within deeply embedded histories.

In this introductory chapter, we distinguish our voices (our stories) from each other's as separate accounts. In subsequent chapters, we knit our stories together as a joint editorial reflection. Although it may be possible for you as readers to discern aspects of our individual narratives, as editors we share the responsibility of framing each chapter in our conceptual framework: transgressive practices to promote equity in every STEM environment possible.

Joy's Story

I am a veteran STEM educator who grew up wanting to be a dancer. Like many of the urban youth (and teachers) described in research reports about science education programs in urban spaces, my Monday through Friday experiences in science classrooms positioned me as disadvantaged with few if any expectations for success. Rodriguez' (1997) notions of standards-induced invisibility and Tate's (2001) thesis on STEM education as a civil right defined my early career as a teacher and helped me understand my experiences as a student in urban and suburban schools. Very little from my home experience was integrated into the curriculum or treated as capital. I was tracked into low-level classes until my first Black science teacher (Mrs. Maxwell), a high school biology teacher in my 10th grade, "peeped my card" and recommended me for honor's chemistry where Ms. Tyson (another Black woman) would be my teacher. I felt betrayed in that moment because I thought my teacher "liked me" and wanted to see me again next year—she did not teach in the honors track—but I suspect she knew Ms. Tyson would make sure I was ready for the 21st century realities of STEM competitiveness initiatives. It was 1985. By 11th grade year, I was fully enrolled and participating, in an engaged way, in accelerated academic programs. Pre-college programs at the city university were opened to me. The "opportunity to learn" concept first took shape for me then ... even though I am only realizing that now.

I graduated one of three chemistry majors that year from an HBCU. There, for only the second time in my life, I was given the opportunity to be autonomous in a functional research lab. The first time was as a summer student intern at a smaller, local pharmaceutical company (thanks to Inroads), this after being rejected from a position at the major, larger local pharmaceutical company because my 17-year old presentation of self seemed aggressive to my interviewer. Really?! I will never forget the mentorship and validation provided by Dr. Floie Vane then and the various others along the way who have helped me to understand why equitable STEM education is a necessary goal for us to pursue as a global community.

Janelle's Story

I entered the teaching field as an English as a Second Language (ESL) educator in 1997. The second language acquisition methods I had learned largely centered on using hands on math and science-based activities. Language acquisition is much more effective when it is connected with meaning-making, and students tend to enjoy hands-on work, which means they tend to be more motivated as well. My first teaching position was at a private school in Guatemala City, teaching elementary math and science in English with Guatemalan students—it was a dual language school. That same year I was asked by my teaching partner to begin offering hands-on science workshops for local teachers through the International Reading Association. I moved from third grade to fourth grade with my students, and the school asked me to teach 7th and 8th grade science in my third year there. I moved back to the U.S. for a brief period, but then returned to Guatemala for another four years to teach at an international school. I was initially hired to teach middle school math and science, but when I arrived was asked to teach some high school math as well. I continued to offer teacher workshops in both math and science that incorporated children's and youth literature, hands-on learning, creating safe spaces for students to learn, and connecting with student interests.

When I started graduate school there was a cohort of Central American teachers at my university there for a year long capacity-building scholarship. I developed friendships with many of the teachers and heard about the communities they came from. I knew how hard they were working, and how dedicated they were to serving as agents of change in their communities. But did this program work? I started to ask, and then I started to investigate the research literature, but I was having trouble finding anything other than reports. I wanted to learn about the outcomes in the teachers' communities, not just documentation of what the program had done and how many teachers had gone through the program. I wanted to hear, from the teachers' perspectives, what they were able to take away from this program. What was successful? What were the barriers? And where were the tensions with the organizations from outside of their communities who were funding the effort to improve schooling in these marginalized spaces?

The next section of the introduction describes the theoretical framework for the volume in greater detail, laying a foundation to move from theory into the practices described in each chapter.

Equity Theory into Practice: A Review of Key Frameworks

The question "why do we need to know this?" repositions theoretical frameworks as critical components of practical discourse. As editors, our use of the term *framework* itself lends credibility to the idea that as researchers and practitioners, we want to build something. We provide an index of key conceptual frameworks and cited theories/theorists for ready reference along with the traditional end-of-volume index. We believe that as authors and editors purposefully cite and give credit to a greater diversity of scholars working to centralize issues of equity in their work, it will be more possible and likely to actually realize the ideals of equity envisioned by diverse scholarship. We are curious about how to design impactful STEM education for all. In the universe of available research about education there are many general frameworks from which we could choose but we are attempting to confirm, extend and refine[23] three: the Aspirations framework, the opportunity-to-learn ideal, and the standards of Kahle's Equity Metric. Describing them in this order allows us to move from general theory into more STEM-focused theories about equitable teaching and learning. We came to this discourse through our curiosities: meeting after a conference on STEM education, we were driven by our doubts to ask new (more questions). The frameworks presented here represent our initial survey of the field.

Aspirations Framework: How Are We Doing?

Commonly associated with the now familiar petitions from education stakeholders about "rigor, relevance and relationships"[24] in schools, the Aspirations framework has driven whole reform movements[25] in education. Infusing agency through social action and imagination into the daily experience of children is ideal for all schools. The Quaglia "Aspirations" framework[26] is scaffolded on ideas like self-worth, engagement, acknowledgment of heroes, leadership, responsibility, fun, excitement, curiosity, creativity, and adventure. All these ideas support what could certainly be constructed as a "house of equity" in STEM learning. In the end, children and teachers develop in magnificent ways: they together become efficacious having both individual and collective pathways to success. The conceptual underpinning of this work on transdisciplinary STEM grew out of our work on and reflections about multicultural science education and its visions of culturally responsive pedagogy[27] and Project 2061's "Science for All." Evolving from its 20th century form, equitable STEM education discourse is intersectional; it interrogates the impact of a range of factors operating simultaneously to promote discovery and innovation. We are almost 20 years into the new millennium. How are we doing?

Opportunity-to-Learn: How Do We Address the Challenge of Serving the Underserved?

As part of a strong policy of standards-based instruction for equitable student outcomes, "opportunities to learn" (OTL) theory was constructed to be a critical factor in diverse classrooms. OTL has particular relevance for equitable science teaching in schools that serve students from non-dominant populations. More than access and exposure, OTL is a paradigm of engagement with one's own questions, history and Indigenous knowledges. Guided by a profound pedagogy of care,[28] hierarchies dissolve and all members of a learning ecosystem interact as partners and stakeholders to challenge social, cultural and academic habits and norms while constructing new understandings. STEM ecosystems connect in-school, after school and out-of-school time learning,[29] addressing an important need for students who have limited access to high quality formal and informal educational opportunities.

Limited exposure to STEM learning is confounded by many factors associated with poverty: "lack of access to supplemental educational programs; poor quality schools; low teacher expectations due to bias and racism/classism; low levels of parental education and involvement; cultural and language differences; negative peer influences; and lack of tacit knowledge about higher education."[30] This variation in access to high quality formal STEM education is why many authors refer to the so-called achievement gap as an opportunity gap[31] or underachievement relative to potential.[32] How do we address the challenge of serving the underserved?[33]

Key Questions for Measuring Equity in Science Teaching

Jane Butler Kahle (1998) frames equitable science teaching with the development of a metric to evaluate progress toward equity. Inclusive of many socio-cultural and socio-political markers, equitable science teaching addresses access (resource) gaps, participation gaps and process (product) gaps with the hope of retaining underrepresented populations in science lifestyles in and out of school.[34] She outlines more than twenty indicators of an "equity metric" in the context of systemic school reform which she asserts is not equal—not all schools are applying principles of equitable education equally nor are schools realizing equity equally as they adopt change. Kahle's "equity metric" was later used in a multiple-case study design to measure five middle schools' progress toward achieving equitable systemic reform in mathematics and science. "Two results occurred: various equity issues were identified in the five case studies, and the metric proved efficacious in identifying barriers to or facilitators of equitable reform in the schools."[35] According to the authors, the metric is useful as a tool for self-evaluation and reflection. This element of the metric makes it particularly useful to practitioners who want to measure the quality of programs and improve their own practices. Characterized by five key ideas, equitable programs that are also high quality prioritize standards, expect participants to be consistently engaged and accept evolution/adaptive change over time. These notions include (1) opportunities to achieve within the context of high standards; (2) equitable resource allocation without regard to differences between working groups or individuals; (3) facilitated participation; (4) access to resources (human, technology, equipment, facilities, curricular); and (5) put-in-place policies or procedures that channel resources to and compensate for subgroup differences. In light of this narrowly focused metric, a few final questions emerged for this discussion: do we have sufficient tools to measure equity in STEM teaching and learning contexts? Where do we go from here?

We are attempting to shed light on the current field almost twenty years into the new millennium by moving from the very general notion of equity and democracy in education to the intermediate goal of transgressive teaching offered by bell hooks, to the more specific applications unique to STEM. We believe our biggest challenge is the equitable education of Black & English language learners (ELL) from non-dominant cultural groups, students with disabilities and students from "high-needs" schools. We use the term *high-needs* to refer to the challenging contexts common in schools where poverty rates are high, resources are limited and there is a profoundly disabling incidence of social, academic, and public-health crises that impact school-related programs.[36]

We believe that STEM learners all along the education continuum from pre-school exposure through professional engagement must be able to hear their voice and their histories (their communities' own stories) as they learn. They must be supported as they attempt to make sense of their experiences; they must feel free to challenge constraining thoughts and ideas from the past. That is how they will build their own understandings while also becoming more efficacious and confident. Preparing teachers for these contexts is an absolute necessity. Limited opportunities to develop esteem as a STEM actor, or to see oneself as having an identity remotely related to the creative enterprises of STEM is a major issue in need of resolution for both teachers and students. This need speaks to the heart of transgressive practice; in order to develop a critical lens on STEM in societal and social contexts, we must render problematic our "good" works, our words, and ideas. With no endpoint or limits, we must weave bits and pieces of highly complex stories together to paint a picture of equity in STEM education. This is our responsibility as researchers and practitioners.

Teacher educators will need to develop new sets of skills in order to prepare teachers who will in turn develop new skills in students. Learning to be teachers that act within the ethics of care who can validate students to trust the learning process is itself a critical skill. Helping students to understand the value of tests and failing is part of this skill set. Helping teachers to be comfortable in the co-generative work of knowledge construction is part of this skill set. Helping schools balance anxiety and threats associated with students' intellectual and technical prowess is part of this skill set. Cultivating care in the classroom is part of this skill set: caring about human spirit and enterprise, play and practice, challenge and change (as much if not more than content) is part of this skill set. Knowing that caring teachers will never be replaced by technology is part of this skill set. We can never know exactly what students will be expected to do in twenty years; we have crafted a thirteen-year 'agreement' in formal education from pre-kindergarten through twelfth grade that we hope gets them close. STEM education policy creates and then oversees this contract with students and their families/communities for at least this thirteen years period of compulsory education; neither the policies nor the contracts have been very precise in their diagnosis of or treatments for STEM education problems. What we do know about human learners is that they all need to interact with others in order to be healthy.[37] We know they will need to be able to think and solve problems that we don't currently know exist. There is much that is uncertain in our knowledge base except that 'we don't know what we don't know,' so we must imagine and form then invent, at times even reinvent or reform. This is the root of human enterprise and industry.

Organization of the Book

This volume is a collection of stories with data and theory in mind. The research articles included in this volume embody and/or extend the critical ideas presented above. The volume is organized into three sections:

(1) Standards and Pedagogy: Applying Research to Practice
(2) Engagement: Extended Learning Opportunities
(3) Transformation: Transgressive Practices Along the Journey

Moving further beyond our initial curiosities, two focal questions guided our review of submissions for this volume:

1. How does this STEM experience (teaching and learning) promote transgressive thinking?
2. What do we need to do differently as STEM education providers, as teaching and learning facilitators and as STEM education researchers?

Embracing the revolutionary nature of transformative research methodologies[38] we expect there to be a dramatic shift in research paradigms and disciplinary silos once the research presented here is folded into the toolkits of researchers and practitioners alike. The research presented here is generally interdisciplinary and often times exploratory. Several chapters are authored by graduate students whose ideas extend those of the mentors that supervise them, but some transcend. Providing a space for reviewed publication of graduate research has been both challenging and rewarding, the evidence of which shines through in our reflections. Closing thoughts and discussion questions follow each chapter and draw attention to inclusive STEM pedagogy. Many well-intentioned education stakeholders are searching for "best practices," but can only apply them in highly prescriptive ways that are often a mismatch for the communities they hope to serve. We understand this penchant, and yet seek to provide support only in this: helping practitioners at every level of STEM education recognize disciplinary core ideas (content), science and engineering practices, and crosscutting concepts (habits of mind) that support equity. The concluding chapter of the volume then plaits the three strands together: our stories, each research narrative, and our hopes for broad execution of equitable STEM. We are trying to capture and share promising and inclusive practices as we grapple with difficult questions in STEM education. This is just the beginning.

Notes

1. Current discussion of bitcoins, virtual currency, is a prime example. See https://www.nytimes.com/2017/10/01/technology/what-is-bitcoin-price.html?action=click&contentCollection=Economy&module=RelatedCoverage&region=EndOfArticle&pgtype=article for more information.
2. There is a broad range of standardized testing that applies here. In addition to content tests for student learning in mathematics and reading for example, behavioral tests designed to measure social-emotional needs and status are being used even though they may be flawed, and mask trauma related to risks and threats that learners have endured. Metrics used to assign mental health and psychological stability ratings to students are biased and filled with errors by the standards of research, as any other statistical product is.

3. The term "former" refers to various times when the establishment of democracy in education was seen as progressive philosophy. Proposed by John Locke in the 17th century, Jean-Jacques Rousseau in the 18th, Leo Tolstoy in the 19th, and John Dewey in the 20th, historically, many philosophers have suggested that formal education is a right (or an entitlement) of the people that should be guaranteed simply because it allows adults to transfer hopes, knowledge and expectations to their young.
4. Studies that examine the relationship between social skills, play and academic achievement are abundant in early childhood literatures; additional insights distilled by experts as "opinion" share many of the same views. See http://www.cnn.com/2010/OPINION/12/29/christakis.play.children.learning/index.html
5. See O'Neil, 2016.
6. Predictive policing exposes a potentially biased application of data use. Based on historical data about crime and arrests, models are generated to make patrol assignments, potentially resulting in over-policing of certain communities. See http://fairness.haverford.edu/ for a full discussion.
7. Feldman, Friedle, Moeller, Scheidegger, & Venkatasubramanian, 2015.
8. Ibid.
9. See Gutiérrez, 2002.
10. Scholars Eddie S. Glaude, Jr. and Marc L. Hill both discuss the "value" gap between African Americans and other groups in the sociopolitical consciousness of others. In both their works, they challenge us to change our perception of the bodies we see as Black or brown (Hill, 2016) and the systems that keep them oppressed (Glaude, 2016) including a change in what we think "matters" as citizens of the United States.
11. "Regular" could mean general education, successful/high performing, middle and upper middle class; able-bodied; White/Asian or any other marker of normalized and dominant cultural groups.
12. Gutiérrez & Rogoff, 2003; Next Generation Science Standards Appendix D, 2013.
13. Moll, Amanti, Neff, & Gonzalez, 1992.
14. González, Moll, & Amanti, 2006; Moll et al., 1992.
15. Deloria & Wildcat, 2001.
16. Ruiz, 1984.
17. During a search for the phrase we came across an education leadership blog well worth subscribing to by Teresa Cole called the "Elastic Scholastic" which highlights aspects of educational leadership for gifted and talented education. See https://theelasticscholastic.wordpress.com/author/tacole79/ for more details.
18. This notion of "essential" is similar to that proposed by McTighe and Wiggins in their work with the Association for Supervision and Curriculum Development around essential questions. For additional insight, see McTighe & Wiggins, 2013.
19. hooks, 1994, p. 129.
20. Sultana, 2007.
21. Boyd, Crowson, & van Geel, 1994.
22. These include capitalizing identities such as Black, White, and Indigenous; using the term Latinx in lieu of Latino or Hispanic and reducing or eliminating gender pronouns when plausible. When reflecting our own voices, we used "I," "our,", and "we" to distinguish

editorial voice and thought from the greater narrative. We also tried to remove phrasing that equated "students of color," "poor," and "low achieving" as if they were synonymous terms.
23. CEROW principle: Confirm, extend and refine others' work.
24. McNulty & Quaglia, 2007.
25. Oberman, 2007.
26. See resources available at http://quagliainstitute.org/qisva/framework/
27. The INTIME project of the University of Northern Iowa provides an overview of culturally responsive teaching & the work of James Bank online.
28. Beck & Cassidy, 2009; Pantazidou & Nair, 1999.
29. See Krishnamurthi et al., 2013.
30. Olszewski-Kubilius, 2006, p. 28.
31. Flores, 2007.
32. Ford & Moore, 2013.
33. Additional perspectives on assessment are available through Gummer & Champagne, 2004; Stevens, 1996.
34. Kahle, 1998a, 1998b.
35. Kahle & Kelly, 2001, p. 79.
36. Klar & Brewer, 2013; High-needs schools can exist in vastly different geographic locations whether in rural, urban or suburban communities.
37. David Ross (2018), CEO of the Partnership for 21st Century Learning, wrote a compelling argument that discusses why communication, collaboration, creativity and critical thinking will form the foundation for interactions between humans and artificially intelligent entities. Building from the core perspectives of the umbrella organization, it would be hard to argue against these claims—at least for now.
38. National Science Board, 2007.

References

Beck, K., & Cassidy, W. (2009). Embedding the ethic of care in school policies and practices. In K. te Riele (Ed.), *Making schools different: Alternative approaches to educating young people* (pp. 55–64). Thousand Oaks, CA: Sage Publications Incorporated.

Boyd, W. L., Crowson, R. L., & van Geel, T. (1994). Rational choice theory and the politics of education: Promise and limitations, *Journal of Education Policy, 9*(5), 127–145.

Deloria, V., Jr., & Wildcat, D. R. (2001). *Power and place: Indian education in America.* Golden, CO: Fulcrum Publishing.

Feldman, M., Friedle, S. A., Moeller, J., Scheidegger, C., & Venkatasubramanian, S. (2015). *Certifying and removing disparate impact.* In Proceedings of the 21st ACM SIGKDD International Conference on Knowledge Discovery and Data Mining (pp. 259–268). Retrieved from https://arxiv.org/pdf/1412.3756.pdf

Flores, A. (2007). Examining disparities in mathematics education: Achievement gap or opportunity gap? *The High School Journal, 91*(1), 29–42.

Ford, D. Y., & Moore, J. L. (2013). Understanding and reversing underachievement, low achievement, and achievement gaps among high-ability African American males in urban school contexts. *The Urban Review, 45*(4), 399–415.

Giroux, H. A., & McLaren, P. (Eds.). (1993). *Between borders: Pedagogy and the politics of cultural studies*. New York: Routledge.

Glaude, E. S., Jr. (2016). *Democracy in Black: How race still enslaves the American soul*. New York: Crown Publishers.

González, N., Moll, L. C., & Amanti, C. (Eds.). (2006). *Funds of knowledge: Theorizing practices in households, communities, and classrooms*. New York: Routledge.

Gummer, E., & Champagne, A. (2004). Classroom assessment of opportunity to learn science through inquiry. In L. Flick & N. Lederman (Eds.), *Scientific inquiry and nature of science* (pp. 263–297). Dordrecht, the Netherlands: Kluwer.

Gutiérrez, R. (2002). Enabling the practice of mathematics teachers in context: Toward a new equity research agenda. *Mathematical Thinking and Learning, 4*(2–3), 145–187.

Gutiérrez, K. D., & Rogoff, B. (2003). Cultural ways of learning: Individual traits or repertoires of practice. *Educational Researcher, 32*(5), 19–25.

Hill, M. L. (2016). *Nobody: Casualties of America's war on the vulnerable from Ferguson to Flint and beyond*. New York: Atria Books.

hooks, b. (1994). *Teaching to transgress*. New York: Routledge.

INTIME. (2002). *Culturally responsive teaching*. Retrieved from https://intime.uni.edu/multicultural-education-introduction

Kahle, J. B. (1998a). Equitable systemic reform in science and mathematics: Assessing progress. *Journal of Women and Minorities in Science and Engineering, 4*(2 & 3), 91–112.

Kahle, J. B. (1998b). *NISE brief: Measuring progress toward equity in science and mathematics education* (NISE Brief Vol. 2, No. 3). Madison, WI: National Institute on Science Education. Retrieved from http://www.wcer.wisc.edu/archive/nise/publications/Briefs/Vol_2_No_3/Vol.2,No3.pdf

Kahle, J. B., & Kelly, M. K. (2001). Equity in reform: Case studies of five middle schools involved in systemic reform. *Journal of Women and Minorities in Science and Engineering, 7*(2), 79–96.

Kahle, J. B., Meece, J., & Scantlebury, K. C. (2000). Urban African American middle school science students: Does standards-based teaching make a difference? *Journal of Research in Science Teaching, 37*(9), 1019–1041.

Klar, H. W., & Brewer, C. A. (2013). Successful leadership in high-needs schools: An examination of core leadership practices enacted in challenging contexts. *Educational Administration Quarterly, 49*(5), 768–808.

Krishnamurthi, A., Bevan, B., Rinehart, J., & Coulon, V. R. (2013). What afterschool STEM does best: How stakeholders describe youth learning outcomes. *Afterschool Matters, 18*, 42–49.

McNulty, R. J., & Quaglia, R. J. (2007). Rigor, relevance and relationships. *School Administrator, 64*(8), 18–23.

McTighe, J., & Wiggins, G. (2013). *Essential questions: Opening doors to student understanding*. Alexandria, VA: ASCD.

Moll, L. C., Amanti, C., Neff, D., & Gonzalez, N. (1992). Funds of knowledge for teaching: Using a qualitative approach to connect homes and classrooms. *Theory into Practice, 31*(2), 132–141.

National Research Council. (2013). *Next generation science standards: For states, by states.* Washington, DC: National Academies Press.

National Research Council. (2012). *A framework for K-12 science education: Practices, crosscutting concepts, and core ideas.* Washington, DC: National Academies Press.

National Science Board. (2007). *Enhancing support of transformative research at the National Science Foundation.* Retrieved from https://www.nsf.gov/nsb/documents/2007/tr_report.pdf

Oberman, I. (2007, April). *Learning from Rudolf Steiner: The relevance of Waldorf education for urban public school reform.* Online Submission. Paper presented at the Annual Meeting of the American Educational Research Association, Chicago, IL. (ERIC Document Reproduction Service No. ED498362)

Olszewski-Kubilius, P. (2006). Addressing the achievement gap between minority and non-minority children: Increasing access and achievement through project EXCITE. *Gifted Child Today, 29*(2), 28–37.

O'Neil, C. (2016). *Weapons of math destruction: How big data increases inequality and threatens democracy.* New York: Crown Publishing.

Pantazidou, M., & Nair, I. (1999). Ethic of care: Guiding principles for engineering teaching & practice. *Journal of Engineering Education, 88*(2), 205–212.

Project 2061/American Association for the Advancement of Science. (1989/2009). *Science for all Americans: A Project 2061 report on literacy goals in science, mathematics, and technology* (Vol. 89). American Association for the Advancement of Science. Retrieved online http://www.project2061.org/publications/bsl/default.htm

Rodriguez, A. J. (1997). The dangerous discourse of invisibility: A critique of the National Research Council's national science education standards. *Journal of Research in Science Teaching, 34*(1), 19–37.

Ross, D. (2018, March 4). *Why the four Cs will become the foundation of human–AI interface.* Retrieved from http://www.gettingsmart.com/2018/03/why-the-4cs-will-become-the-foundation-of-human-ai-interface/?utm_campaign=coschedule&utm_source=twitter&utm_medium=Getting_Smart&utm_content=Why%20The%20Four%20Cs%20Will%20Become%20the%20Foundation%20of%20Human-AI%20Interface

Ruiz, R. (1984). Orientations in language planning. *NABE Journal, 8*(2), 15–34.

Stevens, F. I. (1996). *Opportunity to learn science: Connecting research knowledge to classroom practices.* Philadelphia, PA: Temple University, Mid-Atlantic Lab for Student Success.

Sultana, F. (2007). Reflexivity, positionality and participatory ethics: Negotiating fieldwork dilemmas in international research. *ACME: An International Journal for Critical Geographies, 6*(3), 374–385. Retrieved from https://www.acme-journal.org/index.php/acme/article/view/786/645

Tate, W. (2001). Science education as a civil right: Urban schools and opportunity-to-learn considerations. *Journal of Research in Science Teaching, 38*(9), 1015–1028.

SECTION ONE

Standards and Pedagogy

Applying Research to Practice

Section One of the book is on Standards and Pedagogy. Barnes-Johnson's practitioner focus is on policy, history and language used to describe the theoretical underpinnings of equitable STEM education. The chapter is organized in three parts. The first part contextualizes equitable STEM education using a policy lens. Anchored by Johnson-era policies designed to address a growing American commitment to STEM in a climate of change precipitated by the Civil Rights Movement, details about policies from Eisenhower's administration (1953–1961) through our current political context are described. The second part of the chapter picks up post-Civil Rights era discourse about multicultural education and its evolution to equity pedagogy. Notions of teaching and learning for non-dominant groups as articulated by academics (rather than policy makers) are described. The last part of the chapter provides perspective based on the author's own experiences in the field. Articulating action research, Barnes-Johnson pulls each section together and challenges readers to contemplate the current shifts in language and understanding related to equity in science teaching and learning.

Adjapong's work on Hip-Hop pedagogy in STEM brings students' contexts into the science classroom. This research revolves around developing equitable pedagogical practices to provide urban youth, who traditionally have been marginalized in STEM disciplines, with access in STEM education. Research suggests that students from nondominant groups traditionally fall behind their counterparts of less diverse backgrounds in major content areas, including science. In addition, urban students are less likely to be interested in the sciences partially because educators misunderstand the realities and experiences of urban students and as a result they are not able to demonstrate the relevance of science. This chapter suggests utilizing Hip-Hop/youth culture to develop effective and equitable pedagogical approaches to engage urban youth in STEM and to encourage urban students to view the field of STEM as obtainable as they learn science through their culture.

Asante, DeLisi, McKinley and Barnett describe how an urban hydrofarming program draws on social justice issues relevant to youth to drive engagement and learning. They describe a curricular framework that (1) engages students in a range of interdisciplinary learning experiences including urban farming, solar energy, coding and robotics throughout their high school years, (2) explores how participation in the social justice driven STEM program impacts youth's beliefs about STEM, the role that STEM has in their lives and how they can use their skills to improve their community.

CHAPTER ONE

21st c LEASE

Language of Equity and Access to STEM Education

JOY BARNES-JOHNSON[1]

Abstract

The title of this chapter is both an acronym[1] and a principle for the present policy discussion. A lease is a contract or agreement between at least two parties—one having authority (agency) and one with less—about a commonly shared interest. This agency creates both tension and opportunity. These two notions, (1) the tension between negotiating entities and (2) the opportunity to "co-" (co-author, co-operate, co-labor) has specific relevance in the 21st century and forms the basis for this chapter. In this analysis of history and policy, I present a perspective on this contract (the initial promises,[2] various trusts and subsequent breaches) as it relates to STEM education in the 100-year shift between Dewey's Democracy in Education (originally published in 1916) and now. Dewey warned us of the dangerous products of schooling without usefulness—egoistic specialists.[3] He recognized that science teaching in particular was only useful if learning involved full engagement of the learner: "a mode of practice" based on accumulated observations and

---

1   Joy Barnes-Johnson, Science Educator
    Princeton (NJ) Public Schools
    joybarnesjohnson@princetonk12.org

self-directed understanding. Describing "science as experience becoming rational,"[4] very early in the establishment of "democratic education" as a moral guideline for subject-matter instruction, Dewey points us toward equitable practices that are constructivist, non-hierarchical, inquiry-based and experiential. He even blurred the lines created by disciplinary boundaries seeing literature and history and art and language as necessary conspirators in the office of intelligence. What a wonderful guide to follow; so, where are we now?

The chapter is organized to show defining characteristics of STEM education in general using specific vocabulary related to equitable practice that has emerged since then. These characteristic clauses reflect understandings (including my own) of equitable STEM teaching and while I am attempting to collate a comprehensive list of ideas and concepts or terms, others will undoubtedly have more to add. I am attempting to problematize these "clauses" and will thus provide emphasis to them as the chapter progresses. Each clause will then be used as a potential point of discussion and reflection at the end of the chapter. In this I expose my limitation, my almost fifty-year experience as a STEM education stakeholder. My position as a researcher-practitioner is an important organizing principle. This chapter interrogates the legacy of language used to convey equitable science teaching and learning in the U.S. context. My experiences as a high school teacher in New Jersey will be used as secondary level scaffolding for the chapter. Hence the purpose of this chapter is the mindful exploration of policy and practices that address equitable science education, its principles and constructs.

Editorial Reflections

Growing up is highly disruptive ... in a good way. This chapter walks the reader through personal history and discovery. I, Joy, grew up within walking distance of an industrial laboratory and museum. I can recall a time when the National Park Service was free, and we were allowed to enjoy the docent's speeches without a parent; on lazy summer afternoons, my siblings and I would happily linger on the lots of old Edison. On the walk through our neighborhood we passed blacktop parks and two-family homes, restaurants and markets, a cemetery and old factory buildings with histories we didn't know. The Black Maria was still on its turning "wheel" and we could only imagine what it must have been like to be working with one of the most notoriously genius thinkers of all time. Pictures of him loomed large but he, and his work, was made accessible to us by his stories. Years later there would be much that I would learn about Edison, his workers, his inventions, the neighborhood and the laws that now cause me to take pause at how STEM history is told.

To find out many years later that Lewis Latimer was one of the original Edison Pioneers was interesting: what must that have been like for the progeny of a former slave? To find out that industrial waste from the plant had been dumped only walking distance from the labs in nearby Montclair, the part where "we" lived (working middle class and Black families) was interesting. To learn that those buildings we passed on the way to the lab had formerly been the work site of the "Radium Girls," a different group of working class people who dedicated their lives to work, many losing their lives to work-related cancers. To find out that many of my childhood neighbors lost their homes to banks or had them taken by imminent domain in less than ten years was interesting— made more so by the declaration of a superfund site[5] not too far away.

The goal of this chapter is to connect STEM advocacy and knowledge to history and land use to policy and invention. All have informed who I am as a STEM teacher. I intentionally seek to give students a vocabulary to communicate macroscopic ideas with the same precision as they explore the nuanced meanings behind their observations. Most of the details of my time exploring Edison are gone except the biggest monuments and the smallest inclusions. I still have an Edison quote placard hanging in my classroom that restores for me the spirit of my childhood spectator wandering on the Edison site. I use it to remind myself and my students about the nature of genius: the lopsided sum of inspiration and perspiration.

Being asked to examine my own identity and cultural norms, I, Janelle, reflected on how I came to be more than just a science researcher, educator and thinker. I look like a Becky.[6] In many ways I AM a Becky. I grew up in an upper middle class, largely White suburb of Phoenix. For some reason I always detested being lumped with the group I was a member of and hated when people would make assumptions about me. Of course, everyone hates that, but the Becky part of me didn't realize that. Growing up I was drawn to people whose identities and cultures were different than mine. My best friend growing up from first grade on was African American. When a boy from Denmark joined our class in second grade, I befriended him instantly. I would teach him English and asked him to teach me Danish. When a neighbor who had lived in Japan moved in across the street, I started taking Japanese lessons; I was 10. When an opportunity to travel to the Soviet Union presented itself, I jumped on it and was selected to take the three-week trip with 45 other high school students from Phoenix. And of course, my parents had the means for these kinds of opportunities to be in reach.

When I accepted my first teaching position in Guatemala. I did not speak Spanish. I had studied German in school and had a bit of experience with Japanese and Russian. However, being aware of trends in the school population, I realized it would be useful to learn Spanish. In the Guatemalan private school where I was hired, I was only allowed to speak English with my students, and many of the foreign teachers at the school spoke little to no Spanish, believe it or not. In the social circles we were welcomed into, the elite Guatemalans were extremely proficient in English. However, I quickly entered other circles. I have played soccer essentially my whole life, and I started playing at lunch and on the weekends with the school maintenance workers. The school's physical education teacher, a Guatemalan woman, said I should come play with her team in the women's national league. It was 1997, and the league had just been founded that year.

I was excited. I got all my brand name gear together and took a taxi to the address she gave me, arriving a bit earlier than the time she had said. The practice was on a military base, and I didn't see anyone else around who looked like they were there for the same reason I was. A while passed, and gradually some women trickled in, wearing blue jeans and tennis shoes, coming by bus straight from their jobs in factories around Guatemala City. No one spoke English. Despite our language barrier, the players and coach were very welcoming, and they took pains to look out for me traveling to and from practices where we had to crawl in through holes in fences, on the field, and traveling to and from exhibition games in towns and villages where girls or women playing soccer was absolutely something new.

During my seven years living in Guatemala as a teacher I played as much soccer as I could. Because of my privileged status as a White woman from the U.S., I was made to feel welcome to travel across vast class barriers in ways no local could. Because I had learned to communicate on the field and in informal social situations with my teammates, I became fairly fluent in Spanish. When I returned to Guatemala and went to Mexico for my dissertation research, nearly all the interviews were in Spanish. I was also able to access and analyze educational policy documents in relation to what was happening in classrooms.

Gaining entry into the sacred spaces of non-dominant culture is a privilege and is necessary for change. We had to grow into our border-crossing habits; our literacy afforded us these opportunities. We are literate in the traditional sense of language and numeric sense but we also possess technical literacies that have been gatekeeping forces in the divide between the privileged and the oppressed. Endowed by higher education, we have political agency, economic/financial capital and global literacy that comes from our understanding of social oppression. This chapter is about the language used to propagate inequity in STEM. At the end of the chapter, new language is presented that challenges power structures in STEM policy and elevates equitable practice as the standard for STEM education.

Introduction

lease[7]
lēs/
noun
noun: lease; plural noun: leases
1. a contract by which one party conveys land, property, services, etc., to another for a specified time, usually in return for a periodic payment.

As their primary goal, all systems of education should seek to be useful, create space for full participation by the broadest range of people, and cultivate opportunities for that participation at vastly different and yet specific levels. These goals are drawn from the nearly 100-year old philosophical works of John Dewey (1859–1952), Jean Piaget (1896–1980), and Lev Vygotsky (1896–1934), vital reformers in the world of education who introduced us to curricula that were active and constructivist, embedded in social context, and centered by the curiosities of learners. The language they used to describe their work is critical to the way that other stakeholders have come to understand and experience it; this language has become the vocabulary of the many contracts and promises made with families who send their children to school that at the end of thirteen years, students will be educated. As research practitioners[8] in the field of science education, teachers have a stake in this. Teachers have a unique opportunity to communicate and define these concepts both large and small, mundane and erudite. In many ways, teachers are the knowledge producers and distributors about specific content. At least, that *was* the case in the days before Web 2.0 and open publication. Now, technology revolutions[9] driven by social media, mobile technology, and widespread broadband internet access provide not only acceptable platforms for publication, but perhaps expected ones. In a current commercial touting the future of artificial intelligence, a major software company uses a Hip-Hop entertainer to define technology—as a powerful tool for advancing possibility, adaptability, capability. Building on the legacy of a now historic and revolutionary ad campaign memorialized online as "Think Different," Youtube,[10] a video-sharing platform, houses the message of STEM revolution for ready replay.

The debt I owe to "crowdsourced" scholarship is significant having been obtained in traditional forms, by watching public broadcasting, via Youtube, and TED Talks, through searches on Twitter, and Google Scholar, by surfing NARST & NSTA listservs, or participant observation in online discussion boards, chats and digital forums: all of these sources have shaped STEM and many other fields over the last thirty years. Crowdsourcing, a term coined to represent the unique behavior of groups to build capital, is a common mechanism used to collectively build action, "intellect,"[11] and understanding—the benefit of which is ultimately dependent on community members' commitment to reporting or retelling of truths with accuracy and integrity. "Open source principles are applied to fields outside of computing,"[12] thereby making production of information an act of STEM-like agency. The whole concept of "peer-reviewed" works in this context is being upended in transgressive ways.

Recognizing the value of the collective, this chapter traces the vernacular history of science education policy and practices that embrace equity in science, technology, engineering and mathematics (STEM) classrooms. It interrogates the discourse. It wonders what people are saying about STEM education that is different and more inclusive than what has been said in the past. It pays critical attention to the powerful gift of words[13] used to describe science teaching and learning for the purpose of inclusivity, collective efficacy and personal "sense-making."[14] Sense-making itself requires stakeholders to notice, support and engage how phenomena are observed and described; the patterns of behavior that grow out of the experience of knowledge formation guide talk and embrace voice. "Consensus demands communication … [in the end,] all communication is educative,"[15] and so in preparing this chapter I present a singular hope: that by understanding this vocabulary, researchers and practitioners would be willing to critique the socio-political dynamics that serve as barriers to change. Important intersectional topics that combine issues of relevance to science, technology, and society[16] (like food in/security, climate change, net neutrality, public health, disaster designed-in-mind spaces, and the neurosciences) are at risk of fading from curriculum discourse without these kinds of transgressive critiques. Thus this chapter presents in a rather circumnavigational way the history and evolution of equitable STEM education. The first part of the chapter presents the language of policies related to equitable STEM. The second part of the chapter presents the language of equitable STEM practices. The last section of the chapter provides a contemporary perspective on the topic as policy and practice.

Contextualizing Equitable STEM Education as Policy

Policies purported to create structures and systems of *accountability* in schools have been in place since before the LBJ era (1963–1969). In an attempt to guarantee high quality education in America's schools, President Lyndon B. Johnson signed into law the Elementary and Secondary Education Act of 1965 (ESEA) with his childhood teacher sitting by his side.[17] Unfortunately, the problems endemic to life in segregated schools persist. The ESEA was introduced ten years after the 1954[18] landmark Brown v. Board of Education case as part of Johnson's "War on Poverty" platform. Since then it has become the primary legal basis upon which the United States enacts education reforms. This legislation has undergone multiple and varied changes since 1965, including the No Child Left Behind (NCLB) Act of 2001, and more recently, the Every Student Succeeds Act (ESSA) of 2015.[19] Identification of four accountability groups—populations of economically disadvantaged students, non-dominant racial and ethnic groups, linguistically diverse (Limited English proficient) students, and students with disabilities—is a mainstay of the nation's education policies. Johnson's 1965 "war" is still being fought today, targeting funds and program support for these groups. To these four groups, the federal government has elected to hold itself accountable through a policy framework intended to support equitable education. In addition to school-based funding and support, different types of organizations are acknowledged as educative entities, child safety provisions are made, library resources are gathered, nutrition programs are implemented, and co-curricular and extracurricular activities are planned to meet the demands of equity; these occur in theory, at least. These elements are often proposed components in most bills designed to "amend the ESEA." Surprisingly, very few bills that have been introduced into Congress to address these accountability groups in the context of pre-college science and mathematics education actually become laws. The 111th Congress provide a notable example with the introduction of H.R. 6078, the 21st Century STEM for Girls and Underrepresented Minorities Act of 2010.

Sponsored first in 2010 by California Representative Woolsey, H.R. 6078's provisions included support for stakeholder training outside of school and at every level (student, teacher, parent and administrative), parent education, consumable resources, tutoring, internships, counseling, mentoring and data collection. As of 2018, the bill remains a considerable distance away from becoming law. In four congressional sessions between 2009–2010, the bill has seen less support (as indicated by co-sponsorship) over time. Tied directly to the STEM education "strategic plan" known as the America COMPETES Act [P.L. 110–69], the connection to the ESEA seems only loosely tethered to the original "war on poverty" moor. When the America COMPETES Act was introduced in 2007, it was touted as an "investment" in education and innovation; STEM education was re-cast for global competitiveness in a world economy. While this legislative language made room in public discourse for ideas supportive of college and career readiness, core skills in language arts and mathematics that are universally conceived as utilitarian, and vital technologies for global communication and enterprise, its movement away from "poor people's agency" is hardly ever considered. STEM education for individual gain is not sustainable.

At the dawn of the 21st century, while conceiving STEM education as a basis for increasing competitiveness in a global economy may have *seemed* more inclusive and optimistic, it was simultaneously problematic. Only a few years after the 9/11 attacks on U.S. institutions, in 2007 the social, economic, and political climate in the world was more tumultuous than restorative. Appearing to embrace the best principles of multicultural education—in the form of inclusive school practices, access to resources, continued education, interdisciplinary curricula, engagement with business for authentic experiences—the promised benefits to children were unclear; attention to "gender gaps" and the "digital divide" would be difficult NOT to provide, right? It may be argued, however, that these practices have exacerbated performance gaps between students, especially in critical need content areas like science.[20] Two big ideas related to STEM education and competitiveness were branded under the America COMPETES legislative banner: the notion of STEM as a pathway to global competitiveness and the nationalist idea that in the United States, we have the resources to actually advance STEM at multiple levels. This short title for the "America Creating Opportunities to Meaningfully Promote Excellence in Technology, Education and Science (COMPETES) Act" makes funding provisions for banner stem agencies like NASA, NOAA and NSF; STEM training grants, innovation, green jobs, accountability systems and education (Title X). Title X of the legislation promotes access to advanced classes in mathematics, science and critical foreign languages, however, these programs rarely exist in high-needs communities,[21] instead creating segregated schools within schools. Later and/or related[22] congressional texts build on these principles and perhaps are sustaining current lags in opportunity for underserved communities.

Table 1.1 provides additional insight into this relationship between legislative language, legislative action and the more general purpose for the policies. The America COMPETES legislation (proposed and enacted) executes ESEA's "war on poverty" principles using language that advances meritocracy, entrepreneurial efforts, potentially hypervigilant reporting, consolidation and coordination of resources between agencies, and attention to industries that are limited in the U.S. landscape (e.g., manufacturing, computing and green technology); this is *not* the pathway to equity. According to race indicators in the "Status and Trends in Education" report in the High School Longitudinal Study of 2009,[23] the value of these programs to White and Asian students alone is substantial. I have to hope that these were unintended outcomes as we had hoped to rise above the "gathering storm"[24] of poor performance as a nation, understanding that this legislation was born to address decades of decline in rank among developed nations in STEM performance by various indicators.

The New Jersey Policy Context

In late 2016, amendments to the ESEA took the form of state accountability associated with the Every Student Succeeds Act (ESSA). When ESSA was enacted in late 2015, state educational agencies became aware that student assessment data were not enough to establish or defend a success narrative. Like other states, New Jersey was forthright in implementing the law and continued what I knew to be its transparent nature of making information[25] available to stakeholders. School data

Table 1.1: 21st Century Education Legislation Equity Links to STEM Competitiveness Initiatives.

Legislation	Familiar Titles	Description
H.R. 2272/H.R. 5116 (110th; 2007–2008/111th; 2009–2010)	America Creating Opportunities to Meaningfully Promote Excellence in Technology, Education and Science (COMPETES) Act [P.L. 110–69; P.L.111–358]	Workforce development, including computer programming as a critical foreign language; gender equity
H.R. 2303 (115th; 2017–2018)	America Can Code Act	
S. 1968/H.R. 3316 (115th; 2017–2018)	Code Like a Girl Act	
H.R. 4150 (115th; 2017–2018)	Innovate America Act	Provide grant funds to support secondary and post-secondary STEM education with additional emphasis on informatics (IT), including expansion of Noyce scholarships and incentives for private sector and small business research and development
H.R. 3839 (115th; 2017–2018)	Today's American Dream Act	Investments in multigenerational workforce, community, economic and healthcare development of citizens

Source: Author.

can be found online[26] with very little lag in time or accuracy, which ultimately is a positive outcome of the law.

Diversity in New Jersey provides a unique context for studying problems associated with equity in schools. In 2016, administrative code[27] was publicly released to explain how education agencies should "manage for equality and equity in education." Key definitions of concepts like *achievement gap, affectional or sexual orientation, disability, discriminatory practices, educational equity, multicultural curriculum* and *prejudice* among many other relevant terms provide insight. There is a long history in New Jersey of a policy response to inequity that includes the formation of Abbott districts to address the disparity between children educated in poor, urban school districts. Populated by students requiring "beyond the norm" educational services, Abbott districts are mostly state-run schools with large percentages of poor and minority students[28] that were formed as a result of litigation in the state on behalf of these students. Two major policy strands of this litigation have particular relevance for the present discussion: (1) policies that provide students with early exposure to science and (2) standards-based education. In New Jersey, districts classified as Abbott districts have full-day, state-funded kindergarten for children who enroll, including for children that are three and four years old. This gives students early exposure to formal school settings and multiple opportunities to learn early in their social and cognitive development. State-supported preschool programs began in 1999 in the 31 former Abbott districts of New Jersey as the result of a New Jersey Supreme Court decision. Standards for preschool programs include a maximum class size of 15; certified teachers with early childhood expertise; assistant teachers in every classroom; comprehensive services; and a developmentally appropriate curriculum designed to meet learning standards. These factors have been considered part of the best practices disposition in New Jersey education.

The second area relevant to a discussion of quality science teaching in New Jersey relates to the state's standards movement. In 1996, the State of New Jersey published and widely distributed a science curriculum framework that included teaching resources designed to address the specialized needs of diverse student groups. In the 2004 version of the standards, the authors write "science should be taught at all levels with awareness of its connection to other subjects and the needs of society … The standards also reflect the needs of the students and teachers of New Jersey …" (New Jersey State Department of Education, 2004, p. E-2). Framing the changes to the document as "habits of mind," New Jersey follows principles advanced by national and professional science learning communities to build change. Placing the needs of teachers on par with needs of students makes New Jersey even more interesting from a policy and education research perspective. By 2010, the standards were revised to clarify the teaching standards, but its effects seem to have taken too much away from the documents' original strength, especially for urban schools. As the standards documents evolved, clear differences in the language and structure of the documents become apparent.

Outlining Characteristic "Clauses" of Equitable Science Teaching and Learning: Exploring Theoretical Language

Equitable teaching fits along a continuum of ideas commonly associated with multicultural education. In a comprehensive synthesis of scholarship on multicultural education, Geneva Gay[29] outlines the myriad of ways that multicultural education (MCE) is described. Citing seminal works by James Banks, Christine Sleeter, Carl Grant and others, Gay explains that whether constructed as an idea, a philosophy, a process, a program, a reform or a set of priorities and practices, the definition of multicultural education has evolved since its inception and is progressing toward clarity.

Origins of Multicultural Approaches to Education

Among the most cited works[30] on general multicultural education, the scholarship of James Banks [Approaches to Multicultural Education which were later called Levels of Integration of Multicultural Content (1988/1993a), Dimensions (1993b/1995), Teacher Stages of Ethnicity (1984), Types of Knowledge (1993b)], Christine Sleeter & Carl Grant [Approaches to Multicultural Education (1987/2003)], Geneva Gay [Culturally Responsive Teaching (2000/2010)] and Donna Ford [Multiculturalism for Gifted Education (Six Dimensions of Multicultural Education, 1999; 2001)] is commonly referenced for their emphasis on educational equity and excellence, personal empowerment or attitudes and value clarification particularly for students of color. An element that these works share is their hierarchical organization. In each case a typology is presented that can be used by theorists or practitioners to help organize discourse data—details about interactions between teachers and students within a classroom.

Banksian Multicultural Education

In 1974, James A. Banks set out to define contemporary multiculturalism as its own discourse given the complexities of the changing sociopolitical landscape in the United States. In a keynote address to the Association for Supervision and Curriculum Development (ASCD), he indicated the need to establish "clarifying concepts" in the debate over what multicultural education programs and curricula should be for ethnic minorities enrolled in our nation's schools.[31] In the almost forty years since the need emerged to establish this direction in the discourse, very little has changed in the way minority children perform in schools. An abbreviated view of the research and theoretical legacy James Banks has provided to the field of multicultural education is described in Table 1.2.

Motivated by his interest in addressing the needs of ethnic minority youth living and learning in cities at the end of the Civil Rights Movement in the United States, Banks' perspective was a critical one. Sociology and research traditions that address specific conditions in urban schools and Black students in particular rest on the legacies left by Carter G. Woodson (*Mis-education of the Negro*, 1933), W.E.B. DuBois (*Philadelphia Negro*, 1899; *Souls of Black Folk*, 1903), Kenneth B. Clark

Table 1.2: Banksian Multicultural Education: Short Research Timeline.

Banksian Multicultural Education	Description
The socio-cultural environment of ethnic minority youth[32]	• Bifurcated existence of ethnic youth • School's purpose linked to youth capacity to border cross between sub-cultures
Levels of integration of multicultural content[33]	• Contributions • Additive • Transformative • Social Action
Dimensions of multicultural education[34]	• Content integration • Knowledge construction process • Prejudice reduction • Equity pedagogy • Empowering school culture and social structure

Source: Author.

(*Dark Ghetto: Dilemmas of Social Power*, 1965) and others throughout the late 19th and 20th centuries. These researchers have tried to understand education from within the Black community and have interrogated power structures and hegemonic forces in place to subjugate minorities. The work of these scholars who position the educational experiences of Black students and teachers in the center of their discourse is now classified within the canons of critical race theory or critical race pedagogy.[35] Banks represents for me a late 20th century scholar whose work bridges two centuries. His work continues to shape the research methodologies and theoretical underpinnings of work done to understand the education of Black youth in the same way that the scholarship of Woodson, DuBois and Clark shape the sociological imagination.

Over the span of Banks' research career, the shift in useful terminology to describe/conceptualize the problem becomes clear. His work represents a necessary synthesis of ideas at an almost grass-roots level. Banks defined multicultural education as a viable means of reform and helped scores of educational workers and researchers think about the unique challenges that diversity presented from the perspective of a non-immigrant minority. His work made room for a new lexicon that would come to include terms like "culturally relevant pedagogy" and "equity pedagogy" among the various typologies he outlined with his work. Banks took a typological approach to theorizing that helped practitioners and key stakeholders use the theories being developed in academic institutions. Several models of multicultural science education have been presented that are clearly based on seminal works in Banks-informed multicultural education research. For example, Martin and Atwater[36] used Banks' work to design a typology to identify and articulate characteristics of a multicultural science teacher; they formed a hybrid model that combined much of Banks' work on culture and ethnicity. By creating a checklist of observable characteristics to help delineate who multicultural science teachers are, they armed researchers with tools that could be used to evaluate socio-cultural elements in the development of a competent and qualified science teaching force. Mary Atwater's work[37] helped establish a research agenda for multicultural science education. Based in the tenets of social constructivism (Vygotskyian philosophy) and Banksian multicultural education, multicultural science education is a foundation from which equitable science education has grown.

Equitable Science Teaching

The phrase "equitable science teaching" is relatively new in the language used to describe science education reform. The earliest appearance of the term that I could find among peer-reviewed literature appears well into the 1990s. Researchers at the University of Hawaii describe an in-service model of elementary science teacher training that addresses effective teaching strategies for gender equity.[38] Their results showed positive correlation to student performance and teachers' exposure to training around gender equity. In a different study of novice teachers, gender equity and collaboration were advanced as a basis for professional knowledge standards in science teacher education.[39]

By 2001, evidence of a research agenda around equitable science teaching was apparent in the Journal of Research in Science Teaching. In volume 38, the first of several publications co-edited by Angela Calabrese Barton, a leader in urban science education research, issues of equity in the urban school context were explored in depth. Equitable systems of science education, which include the teacher, create opportunities for students by providing resources, access and opportunities to learn quality science without regard to class, gender, culture or able-bodiedness/ability allowing for high achievement and retention of students in science throughout their education and life.[40]

CRP

CRP as an acronym has come to mean different things and is used in a variety of ways. *Critical race pedagogy*[41] is a collective theoretical response by scholars to describe the disparate outcomes of students from marginalized groups. Using legal and sociology studies as the basis for its description, critical race pedagogues like Derrick Bell and Marvin Lynn are cited by many of the authors outlined above. A particularly salient view of equitable science teaching is provided by at least one example that draws from literature about culturally *specific*[42] pedagogy. The state of

Alaska has adopted *culturally relevant/responsive pedagogy* as a model for all of its instruction and teacher preparation. In the Indigenous communities of Alaska,[43] cultural responsiveness is critically important to education policy makers. Published first in 2001, the Alaska Science Consortium unapologetically engages spiritual values as a *diversity text* in the science classroom, making available a handbook[44] for its inclusion in the science curriculum. Involving local experts and valuing student beliefs and skills are just two of the characteristics described by Sidney Stephens, parallel to earlier views around opportunities to learn science. At present I propose we may be on the cusp of a new language shift engaging terms that move the discourse away from the original theories of critical race and culturally relevant/responsive pedagogy into a new idea: *community relevant/responsive pedagogy*. The latter view considers a model of shared knowledge building rather than knowledge transfer and views stakeholders as contributors to the knowledge building process.

> The role of the school is not to ignore or replace prior understanding but to recognize and make connections to that understanding ... The culturally responsive curriculum [advanced in the school then] devotes substantial blocks of time and provides ample opportunity for students to develop a deeper understanding of culturally significant knowledge linked to science. (Stephens, 2001, p. 7)

Opportunity to Learn (OTL) theory has particular relevance for standards-based, equitable science teaching in an urban school; when instruction is designed well, delivered with great intentionality and exposes content for personal meaning-making, students have adequate opportunities to learn.[45] There are other key examples where OTL is defined using culture as a key "starting point,"[46] providing STEM fields with an important theoretical paradigm for education: social justice. Social justice as "the elimination of *institution*-alized (emphasis added) domination and oppression"[47] is an ideal goal for schooling. Schools operationalize the responsibility to treat everyone according to their individual needs while also working to treat all groups the same by creating tracks for students—every child has a track;[48] unfortunately, not all tracks will give every child an equal opportunity to learn and not all efforts to develop equal programs can deliver equitable results. Perhaps among the barriers to equitable education are invisibility and neutrality.[49] How could programs be designed to mitigate these factors? Science-technology-society curricula were designed to address these problems creating opportunities for students to learn science in ways that contextualized human needs. The movement toward "for all" education was no doubt assisted by the legacy of a strong institutional memory embedded in the work done by the American Association for the Advancement of Science. A short review of "for all" science education history will help make that point clear.

"Science for All:" A Historical Movement of Science Education Reform

In 1922, Sir Richard Gregory, long-standing editor of *Nature* magazine and president of the British Association for the Advancement of Science wrote that

> ... in science the most appropriate instruction for a class as an entity must be that which expands the vision and creates a spirit of reverence for nature and the poser of [humanity] and not that which aims solely at training scientific investigators.[50] (p. 438)

Gregory realized the importance of cultivating student interest and creating opportunities for school communities to build agency for the ordinary things that have a base in science habits of mind rather than science disciplines. Perusal of his writings and the catalogues available for Science reveals that "for all" approaches to science teaching have been around in the United States and Europe since the early 1900s. When the American Association for the Advancement of Science published *Project 2061: Science for All Americans* in 1989, they were in effect maintaining a view that they had held as an organization for many years.

In 1993, the AAAS published *Benchmarks for Science Literacy* as a follow up to the original work. This latter work transformed industry ideals into goals for K-12 schools. Since then, two publications have served as guides for standards-based education reform efforts in the United States and world, *Benchmarks*[51] (AAAS, 2009) and the National Research Council's (1996) *Science Education Standards*. Three principles are advanced by this movement: a principle of inquiry in science, a principle of equivocal achievement through science literacy, and a principle of equity—parallel participation—in science. Formal science learning tends to start in elementary schools where "for all science" teaching and learning is a standard for general education. The National Science Teacher Association has centralized "Science for All" for the science educator since the mid-1990s. In its *Science Teacher* publication, designed with a high school teaching audience in mind, every volume has at least one dedicated issue for equity education framed within the "for all" construct. Because gender equity in STEM research is widely available, frameworks exist that can be applied relatively easily to classrooms. Published by the National Science Teachers Association in 2008, "the triad framework" to address *Girls in Science* examines the interactions between student goals, teaching goals and science goals to define quality science "for all" and "by all."

This framework provides an action plan and rubric for key stakeholders to follow as they build equitable science learning experiences for all students, not just girls. Simple things like using wait time, involving everyone in the feedback process, allowing students to touch and feel and do, learning to see inequity and continually learning new things and using them to improve teaching are among the most important teaching goals in the triad framework. The idea behind the framework is this: as teachers accept this greater responsibility to increase the decision-making capacity of students, social justice outcomes will emerge. The authors believe that students will become more scientifically literate, critically aware of science uses in their everyday lives, technologically competent, and considerably more able to make informed decisions about their health, the environment and their resources. However, normalizing behaviors of girls could be a problem, especially when considering the social norms of other groups. In 2017, there are so many alternatives to the traditional identity narrative that gender equity frames are no longer sufficient to describe inequities. The "for all" notion has been problematized by many in the field, especially for "non-mainstream"[52] students: urban students, linguistically diverse populations and geographically isolated students.

Urban Schools

Among the many education scholars who have written about urban schools, Martin Haberman[53] (general context) and Rochelle Gutiérrez[54] (in mathematics teaching and learning context) help illuminate some of the reasons why urban STEM pedagogy must be approached differently than in other contexts. Five key issues listed here represent a synthesis of those ideas along with my own experiences in urban and suburban schools. Each has informed my understanding of the long-standing debate over STEM education in urban schools: (1) disparaging impact of accountability policies (standards and assessment) on science instruction in low performing schools; (2) multiple indicators of low attainment among urban students in science and mathematics; (3) low incidence of engaging (non-textbook based) science instruction at the elementary level; (4) consistent and sustained under-representation of specific groups in STEM, especially low SES, special needs and non-dominant racial/ethnic minority populations and (5) low science teaching efficacy beliefs of urban elementary teachers.

In urban elementary school settings, science consistently is among the lowest resourced content area in terms of instructional time, highly qualified teacher distribution, curriculum resources, technology or equipment.[55] Teachers in urban classrooms often find themselves on the 'front line' in the battle to improve science education in the United States with few resources, or perhaps worse, resources they are untrained to use. Further complicating the issue is that many minority-serving communities find it difficult to embrace "for-all" initiatives[56] that are often formulaic or at odds with their belief systems. Students generally love science but "hate" *transmissive methods*[57] used to teach it in school,[58] contributing to students' and families' lack of enthusiasm for school-based science.

The high-stakes testing environments that typify the present school-society structure places pressure on teachers to incorporate more language arts and mathematics instruction into the school results in limits on science instruction. Schools located in impoverished communities where funding is calculated based on test performance data are the most likely group to be represented in schools like those characterized by McMurrer.[59] Commonly referred to as "high-needs" schools[60] or "high-poverty" schools,[61] many urban schools are characterized by limits and constraints on a personal/family and community level; these schools are often times the site of accountability conditions that require several rounds of high stakes testing as part of their funding formula. The conditions in *high-needs* schools often shape teacher expectations. In the science classroom, students and teachers interact in ways that may either build understanding or contribute to misconceptions; students will likely either develop skills and habits that support lifelong learning and capacity building or develop attitudes and beliefs about science that harbor fear and ignorance. Conceptualized as "negative positioning,"[62] low expectations of teachers for specific groups is damaging for student development. The choices[63] teachers make in support of equitable practices are important considerations when thinking about teacher quality and should be used as a viable metric.

Practitioner's Perspective on Equitable Science Teaching/Learning: A Contemporary View

Many practitioners would like to use findings from good research. Unfortunately, teachers are commonly victimized by research findings and policies that condescend or reinforce the status quo but strong pedagogues are often willing to asking new questions and demand more from themselves. For example, a question posed by the National Association for Research in Science Teaching[64] is appropriate for consideration: *How will all educators and society have a deliberate coming together to envision equity as a guiding framework for the implementation of NGSS?* This is an incredibly useful and reflexive question for thinking about categories of equitable science teaching and learning (EST/L) in light of the dynamic nature of the science classroom.

Dr. Kimberly Tanner of San Francisco State University offers a slightly different perspective on equitable science teaching: hers is a self-assessment[65] of strategic use of twenty-one questions (self-reflections) for teachers. Organized into four categories for biology teaching, this list of twenty-one techniques can be used as a blueprint for building a better, more engaging learning environment for all students in all technical (technically any) classrooms. More recently, members of her research team developed a tool that examines class noise level as an indicator of active learning in the classroom.[66] This work has inspired my own reflection. The chart below represents my internalization of a few of those ideations of theory put into practices.

Table 1.3: Applying Tanner's Equitable Teaching: Responses to Strategic Self-Assessment Questions.

Strategic Question	Theory into Practice (TIP)
Have you given students opportunities to THINK and TALK about the content (DCI)?	Expect the Great! 1. Maintain a carbonless class notebook in real-time 2. Be guided by standards documents as a minimum and a metric

What are you doing to ensure ALL STUDENTS PARTICIPATE?
 	
Does your classroom praxis reflect INCLUSION and FAIRNESS?
	Messaging about community
	1. Day one placards
	  a. Doodle tool
	  b. Name for me, message for them and a near-neighbor
	2. Intentional groupings that are dynamic
	  a. Safety to celebrate new alliances
	  b. Expectation of productivity
	  c. Rapid change is possible
	3. Are we having fun?
	  a. Do we all feel safe?
	  b. Can we all smile every day?
	  c. Is what we are doing relevant to life outside of school?

Can you hear evidence of DIVERGENT THINKING in your and your students' discussion?	Teacher-Behavior modification:
	Use thoughtful wait time silence but affirm non-verbal communication
	  1. Address misconceptions as they become apparent by asking more questions that require clarification
	  2. Accept that as the teacher, you may not know "the answer" and that students' perspectives are shaped by a variety of personal experiences that may be unique to their cultural norms.
	  3. Eliminate "good" when providing feedback to students; this type of value judgment is often meaningless in broad context but build a reward system based on execution of habits of mind supportive of strong communication/collaboration

Is what you do, HIGH QUALITY and MINDFULLY crafted?	7-E lesson planning[67] (Elicit interest, Engage with questions and observations, Explore with new activities, Explain by connecting old understandings/conceptions to new learning, Elaborate for differentiation, Extend by connecting to other content areas, Evaluate learning in formal and informal ways)

Source: Author.

Equitable Science Education Theory into Practice: Personal Perspective

Based on evidence from the kindergarten cohort of the Early Childhood Longitudinal Study (ECLS-K)[68] researchers have posited that a "children's motivational beliefs facilitate their engagement with learning activities, thereby increasing the likelihood of successful learning" (p. 219) so I sought to do action research in an early childhood environment. The research question—*how can science teaching in urban schools be improved to reflect high quality practices*—was informed by the language used in policies that were directly impacting schools. By examining veteran teachers' self-efficacy beliefs[69] about equitable science teaching (EST), I was hoping to provide insight about teachers' actions, students' interests and motivations to do science, and the supports needed for increased sustainable instructional supports for science in high-needs schools. By conducting an investigation that (1) treated the urban setting as capital, (2) looked at the system as a valuable pathway for learning opportunities in science, and (3) expected both students and teachers to achieve success, it was my own hope that I could impact conversations about equity as best practice and make a statement about teacher quality. I wanted to somehow investigate the intersection of teachers' (practitioners') beliefs and actions with theories about equity and multicultural science education. I subscribe(d) to the belief that the latter would lead to student success; stakeholders' own belief systems and knowledge bases could be used as currency to motivate personal learning and success. An underlying assumption in my doctoral work was this: because teachers have themselves generally had a broader range of opportunities to develop their own beliefs and practices, the cumulative effect of challenging urban teaching contexts may be overcome in terms of successful teaching and learning. Drawing from the *funds of knowledge* available through students' own *diversity texts*,[70] teachers could help stimulate rich experiences in the classroom. The research[71] brought to bear many of these ideas.

As a researcher, I delightfully watched as a teacher participant had an aha moment involving the use of music and literature in the classroom, not classical music but popular music—Hip-Hop. A teacher with low self-efficacy for science teaching[72] wanted to do a dissection with her early childhood class by the end of November. We discussed the value of doing a dissection using an egg first, then a chicken wing. She began to understand why doing a whole-body animal dissection was not appropriate. In a very short span of time, she started imagining a way to use the chicken part dissection to position the class for readiness to incubate a chicken embryo for lessons in the spring. She made up a song called "hair, skin, muscles and bones" to help explain to students what they would observe when we did the dissection. She created an ideal context for students to compare their own arm to the chicken wing—building on her established routine of relating new vocabulary and new objects to familiar objects. After talking about this unit plan, she took the lead on the lesson that would use the Hungarian folktale "Stone Soup," a story book by Carmen Lomas Garza called *Family Pictures* (*Cuadros de Familia*), and a Hip-Hop song called "Chicken Noodle Soup." She designed a set of lessons to teach children how to make soup from chicken. Seeing this unit as an opportunity to talk about nutrition and build a coherent lesson around food, the teaching team and instructional aide created a pot of soup using the chicken wings we observed and "dissected" during class, along with locally grown carrots, onions and Sazon![73] The children laughed, sang, and danced their way into learning the purpose of muscles: to move bones. A visit several months later confirmed this knowledge acquisition.

Music in the classroom is a dominant teaching style for promoting equity, especially for Black children.[74] Music can be used as a tool of engagement and socialization that capitalizes on harmony, movement, verve and expressive individuality in the classroom. The preference for visual and aural reception of content over print has been reported for a diversity of ethnic and minority children.[75] These socialization and youth learning styles have been compared and correlated in newer theoretical models of multicultural education for inclusive school settings.[76] Researchers are responding to the need to describe best STEM teaching and learning practices based on observable space and actions of learning stakeholders in a space.

Conclusions and Reflections

What does transformative equitable STEM education look like in the 21st century? In a metasynthesis of fifty-two empirical studies, Julie Brown[77] presents a compelling case for complementarity of cultural responsiveness and inquiry using language of the Next Generation Science Standards' three-dimensional learning. Brown reports that

> ... as they engaged in science experiences that were empowering, validating, and relevant to their lives, students of color, English language learners, and low-income students practice many aspects of inquiry that reflect Western science ... in the form of obtaining, evaluating and communicating information that became the foundation for evidence-based claims made by students about scientific phenomena. (p. 1166)

This assertion is one that reflects research useful to practitioners. Brown goes on to state what many practitioners know about research: teachers need examples, not just "findings." What makes Brown's work so important, however, is the openness and transparency with which she shares her data: the author publishes a digital file⁷³ that details her sources and provides insight into her methods, codes and themes for full review. Practitioners and other researchers can see how the data were synthesized. This is transgressive practice at its best.

Transformative STEM education must engage thinking across multiple disciplines. It goes beyond interdisciplinarity into a level of ascendant embeddedness that erases the boundaries created by disciplinary "silos," a necessary model of border-crossing and transgression for education stakeholders. As an example, in my reflection on a talk⁷⁹ given by artist Titus Kaphar, I was forced to think about who I am as a teacher in a community where scholarship is being born every day just a few minutes' walk away. I am one of few Black teachers. I am a science teacher. I once (as a child) aspired to be a dancer. Like many in the field, my thoughts about STEM education have evolved to become dispositions on STEAM education.⁸⁰ Following the discourse on social media, I smile because researchers and practitioners "banter" about this in so many ways: STEM, STEAM, STREAM,⁸¹ including more and more content areas. This is a discourse of intersectionality/interdisciplinarity. Another related current and exciting trend in STEM education across all strata is the development of *makerspaces* for students. These purposely designed platforms for interaction and engagement are interdisciplinary and constructivist, promoting collaboration and critique seamlessly. Two primary premises of makerspace are "experiential learning (trial-and-error, learning-by-doing, student-centred), and social constructivism (new knowledge is developed through collaboration, social interactions, and the use of language)"⁸² (p. 23).

Our BIGGEST challenges are the equitable education of Black students, ELL students from non-dominant groups, students with disabilities, and students in urban/high-needs schools. Educating teachers and STEM experts for these special contexts is critical. Limited opportunities to develop esteem as a STEM actor, or to see oneself as having an identity remotely related to the creative enterprises of STEM is a major issue in need of resolution that is related to the language we all accept, internalize and advance as normal. This graphic represents what I consider to be a salient history of the policy-into-practice discourse. Because we have adopted the early language we may have contributed to the entrenching forces that have stabilized inequity. Thankfully, as more STEM actors intentionally work to disrupt old systems, we are getting closer to addressing the problems of inequity in STEM and STEM education head on.

Alternative routes to certification were once proposed as a solution in schools, but there is compelling evidence from the Education Commissions of the States (Mikulecky, Shkodriani & Wilner, 2004) and other researchers⁸³ that this is not the answer. I am admittedly disappointed in this conclusion because I am

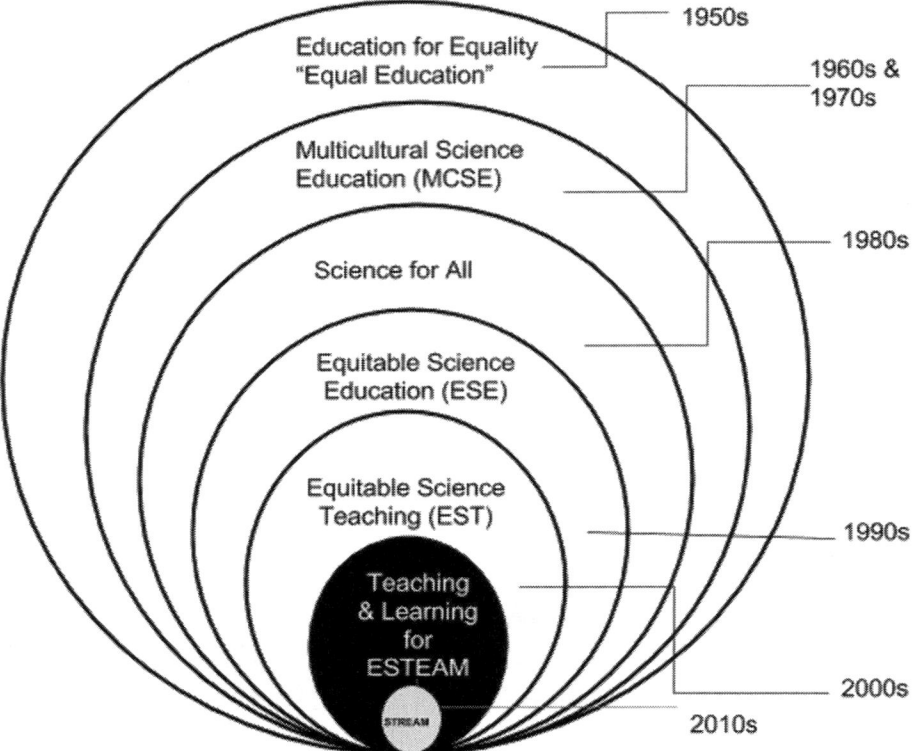

Figure 1.1: Nested History of the Development of Equitable STEM Education Includes Several Shifts in Language. ESE: Equitable Science Education; EST: Equitable Science Teaching; ESTEAM: Teaching and Learning for Equitable Science Technology Engineering Art & Mathematics; STREAM: Science Technology Reading Engineering Art Mathematics. Source: Author.

alternatively certified, so this work is deeply personal. I have been the victim of tracking and have even participated as a teacher within it. Being transgressive is not easy. Many American school reforms throughout history have articulated "education for equity" as a primary goal.[84] The introduction of "trauma-informed praxis"[85] into the discourse about how to teach students (and how to prepare teachers) has created a direct link between schools and the communities they serve. Today's school programs must address the needs and goals of students from families labeled as refugees and/or the children of undocumented workers with the same degree of quality as they do for students from "familiarly diverse" communities based on gender,[86] racial, ethnic, ability, class, and linguistic difference. Equitable science teaching and learning (EST/L) is a construct that provides researchers with a context for understanding the beliefs and practices, the choices and the responses that teachers make in light of the various differences they face in the classroom. EST/L engages both teachers and students in real-time interactivity and grappling. EST/L diminishes power dynamics that position certain knowledges over others while allowing students to develop skills that prepare them for a range of different knowledges. Gutiérrez's (2002) equity framework advances the goal of providing a mechanism for the meaningful contributions of marginalized students to participate in the "erasure of inequity on the planet" as problem finders, solvers and producers of change will require deliberately focused effort. These efforts will likely build stakeholders' confidence, self-efficacy beliefs and esteem characteristics. ESTEAM is a subset of this ideal. The play on words is intentional. ESTEAM is an artistic (creative) pathway to esteem building—a mechanism that leads to James Banks' fifth dimension of multicultural education, empowerment—in science, technology, engineering and mathematics. STREAM is an even more narrowly focused subset of ESTEAM. Fully aware of the power of language to transform students' learning experiences, aspects of literacy (reading, rhyming, [w]riting, righting, and researching)[87] are positioned within the equitable STEAM teaching and learning context. STREAM capitalizes on aspects of performance that honor oral traditions and leverage high-contact[88] cultural norms. These models have a profound impact on theory-to-practice instructional designs. Recognizing that in the margins of childrens' lives, students may experience unique moments of transformation that create interest, define purpose and provide opportunities to ask questions or model STEM best practices, Eq-STREAM teaching for ESTEAM learning is a paradigm to which STEM education stakeholders need to shift. When a STEM teacher is committed to advancing equity in her classroom, she will create experiences for students where multiple literacies in a variety of disciplines emerge as grounds for scientific study. In this environment, students feel empowered to interact with empathy and imagine solutions to complex problems—the best and most transgressive outcomes of learning.

Eq-STREAM Teaching Promotes ESTEAM Learning

Eq-STREAM teaching is an emergent notion of pedagogy that places equity at the helm of the STEM-education narrative. The practices and processes that form the habits of good STEM are enhanced by creativity and imagination (art). These form the basis for teachers' choices, explaining all of the key elements of STREAM teaching except the "R". Representation, the intentional inclusion of multiple perspectives to accommodate both the individual and collective efficacy needs of classroom stakeholders, is centralized in the Eq-STREAM model of teaching. There are two explanatory frameworks for the "R." The first line of thinking around representation and "R" in STREAM can be discussed from a more conventional content perspective: as a mechanism for evaluation using traditional and contemporary tools where students are asked to read, write, rhyme, or conduct research. The more transgressive requirement to ask students to take a "righting" stance or social justice position on issues of STEM is a more transgressive exegesis of the representation idea. The second framework for the "R" considers what it means to be representative or represented in STEM. Expecting students to contemplate others' views, to accurately model information, to cite and provide authorship to marginalized points-of-view, and to take the risk of interrogating assumptions made in data sources (e.g., IndiGenomics)[89] is rather extreme yet transgressive.

It is my belief that in doing teaching this way, the concerns of students from the margins move closer to the center while those from dominant groups develop empathy and understanding that privilege may not otherwise afford them. Evaluation principles of Eq-STREAM teaching rest in three things: 1) the requirement that claims be well substantiated, 2) there be mutual benefit within the community of learners without abuse or exploitation, and 3) that the presentation of ideas and opportunities to demonstrate understanding be executed with significant flexibility; Dr. Nettrice Gaskins, a digital artist, critical pedagogue and STEAM educator is emerging as a leader in this work. These principles lead to ESTEAM learning—empowering science, technology, engineering, art, and mathematics learning. In the Eq-STREAM teaching and learning context, students position themselves to grow and approach traditional learning with clear purposes in mind. These purposes may include personal rewards and recognition but also create a "we win" climate for collective growth and high performance.

Where Are We Today?

The Every Student Succeeds Act (ESSA), our current law, uses "Developing," "Competent" and "Exemplary" to evaluate teachers' and students' achievement. Having more ambiguous words and/or using less precise language to describe the state of education, especially STEM education, is problematic.

> [Science] is more than a school subject, or the periodic table, or the properties of waves. It is an approach to the world, a critical way to understand and explore and engage with the world, and then have the capacity to change that world … (President Barack Obama, March 23, 2015)[90]

I am writing at the end of 2017, a rather confusing year under a newly elected president in the United States. Where we will be at this reading is unclear nevertheless, I write and reflect on a specific trend in STEM education policy emerging right now. Embracing the exciting aspects and references to the former president's comments, I now teach waves with brand new acuity to Hip-Hop culture and technology demands. We "explore the DNA of music"[91] online, bringing scientific, academic, and technical vocabulary into the cultural lexicon of Hip-Hop. As a class we consider "near field communication" (NFC), bluetooth technology, and the differences between those and infrared-remote control technologies—I posed just one question ("what is the difference between controllers for your gaming system, your new speaker for virtual play and my old school remote control?") and opened up a several-days long "battle" about wave technology. The highly competitive character of gaming and sports have added new depth and meaning to discussions of waves. Students feel supported to ask provocative questions about tools needed to track each other in virtual space, details of personalized performance tracking or real-time professional athlete data manipulation for the creation of "fantasy teams" for social league play. The highly technical world associated with informational technology makes sense through youth gaming culture and many of them elect to participate, freely, voluntarily, equitably.

It is nearly impossible to be apolitical in the present context but I hope that it is yet feasible to read this reflection as nonpartisan. When President Obama organized the Committee on STEM Education in 2015, it was his vision to create a coalition (co-authors, co-operators, co-laborers) of "mission-science" and "mission-education" change agents that would be empowered to increase the impact of federal resources. Five focal areas were in his sight:

- STEM teaching and learning in pre-K through 12
- STEM engagement for youth and the public
- STEM experiences for undergraduates
- Increased participation of underrepresented groups in the STEM pipeline
- Graduate education that could support the needs of STEM in the labor and workforce landscape

It took two years to realize this vision as part of public discussion. The plan was crafted originally in 2013. In 2018, it seems to be lost. The link to the "plan" is broken. Shared as a "microsite" from the White House website, an error code now manifests in the place where text formerly appeared. Deeper searches for the work of the committee on the NSF partner website exposes new acronyms in the forms of policy and grants (e.g. "S.T.E.M. for STEM Act" [H.R. 4973], "G.R.I.P." and "A.C.E.S.S.E." (pronounced like access) which stand for Spurring Teacher Education Movement for STEM, Graduate Research Internship Program, and Advancing Coherent and Equitable Systems for Science Education, respectively).[92] Policies that impact science, science education, higher education, higher education funding, scholarships, grants, diversity initiatives and all of the ancillary and programmatic STEM equity progeny that can be imagined are being threatened. At this writing, various media are reporting the existence of a "banned words" list related to important STEM topics under the current administration. Without entertaining the politics of that assertion, I am reflective about the power of words to impact beliefs and understandings. I am also cognizant of the veil placed over policy in the form of legalese and political jargon designed to conceal the real intent of law, threatening access to or blocking channels to resources.

STEM/Education is not value-free, neither are its STEAM progeny or science and mathematics antecedents. In every endeavor to teach or learn, classroom stakeholders (students and teachers) make choices that are short-term, serve self-interests and maximize efficiency.[93] The words and measures we use to communicate STEM speak loudly, especially when standing outside of the margins of widely accepted views or diametrically opposed to personal understandings and cultural norms.[94] When we teach[95] students about three-parent children (thanks to mitochondrial DNA "transplants")[96] and sexual identities (e.g., cis-, trans-, non-binary classifications and multiple identities for individuals) that are distinct from the traditional XX-XY dichotomy, we are opening a world of possibility that can be in direct conflict with religious beliefs and symbols. As an example from the Next Generation Standards (NGSS Lead States, 2013) a biology teacher may choose to emphasize the following list of standards in a genetics unit where students explore their own questions.

Students who demonstrate understanding can:

- [HS-LS3-1] Ask questions to clarify relationships about the role of DNA and chromosomes in coding the instructions for characteristic traits passed from parents to offspring.
- [HS-LS3-2] Make and defend a claim based on evidence that inheritable genetic variations may result from (1) new genetic combinations through meiosis, (2) viable errors occurring during replication, and/or (3) mutations caused by environmental factors.
- [HS-LS4-2] Construct an explanation based on evidence that the process of evolution primarily results from four factors: (1) the potential for a species to increase in number, (2) the heritable genetic variation of individuals in a species due to mutation and sexual reproduction, (3) competition for

limited resources, and (4) the proliferation of those organisms that are better able to survive and reproduce in the environment.

The question "why do we need to know this?" repositions the rigor/relevance/relationship and "aspirations" frameworks as critical components of the discourse. This is the heart of transgressive practice. In order to develop a critical lens on STEM in societal and social context, we must render problematic our "good" works, our words, and ideas. With no endpoint or limits, we must weave bits and pieces of highly complex stories together to paint a picture of equity in STEM education. This is our responsibility as researchers AND practitioners.

Table 1.4: Highlighted Chapter Vocabulary (alphabetical).

Accountability	Diversity Texts	Framework	Learning Progression
Constructivist	Engagement	Funds of Knowledge	Makerspace(s)
CRP	Experiential	Institution	Transmissive Methods

Source: Author.

Discussion Questions

- Borrowing from typical K-12 textbook formatting, several terms are highlighted within the body of the chapter. How would you problematize the highlighted vocabulary in your teaching/learning?
- Identify a teaching or learning standard that you are currently presenting in traditional ways. How could you transform the teaching so that learning is more equitable?
- Follow a line of discussion about science teaching or learning on social media. What language is being used to support equity (or inequity)? Who contributes most to the discussion? Why?
- Examine your own identities and cultural norms outside of who you are as a science researcher, educator or thinker. When do these identities and norms cause a conflict for you or others in your work environment? What are the patterns for resolving these conflicts?
- Explore the distinctions in language used to create paradigm shifts in education being advanced in the chapter. Examples from the chapter include a) the change from critical race, to culturally relevant, to culturally responsive to community relevant/responsive pedagogy and b) the evolution of equitable science teaching and learning to Eq-STREAM teaching and ESTEAM learning.

Notes

1. Use of acronyms, like any jargon, may be a culturally normative type of coding; in the same way that vernacular speech is used; acronyms are often crafted with multiple meanings understood most by those with insider perspectives.
2. Although beyond the scope of the present paper, the Keep Our PACT ACT bill [H.R. 3581] uses this language in reference to support of programs that consider both the ESEA and the Individuals with Disability Education (IDEA) Act.
3. Democracy and Education, p. 9.
4. Ibid., p. 262.
5. Visit https://www.epa.gov/superfund/search-superfund-sites-where-you-live for additional information about these and other EPA monitored locations in the country.
6. Examples in Hip-Hop of "Becky" reference songs include those by Sir Mix-A-Lot, and Jay-Z. USA Today provides a "history" of the phrase in popular culture after Beyonce's reference in the visual album "Lemonade". See https://www.usatoday.com/story/life/entertainthis/2016/04/27/what-does-becky-mean-heres-history-behind-beyoncs-lemonade-lyric-sparked-firestorm/83555996/
7. Definition provided by Oxford Dictionary online retrieved from https://en.oxforddictionaries.com/definition/lease
8. Teachers are enactors of research. In voluntary and involuntary ways, teachers often implement practices to varying degrees based on research. Once elevated to the level of "master" through evaluation and time, veteran teachers often do action research and can be deemed scholar-thinking-doers, a phrase I use to describe those of us who choose to inform our practice by current literature to a greater extent than prior experience. See Landrum, Cook, Tankersley, & Fitzgerald, 2007.
9. Pew Research Center has studied digital technology revolutions over several years and has documented how technology is being used to consume and create novel information. See http://www.pewinternet.org/three-technology-revolutions/ for more insight.
10. See https://www.youtube.com/watch?v=9tucY7Jhhs4 for a close-caption version of the current technology revolution advertisement or https://www.youtube.com/watch?v=c-FEarBzelBs for the historic one.
11. Brabham, 2009.
12. Howe, 2006.
13. "The gift of words" Ted Talk by Javed Akhtar is retrieved from https://www.ted.com/talks/javed_akhtar_the_gift_of_words
14. Bang, Brown, Barton, Roseberry, & Warren, 2017.
15. Dewey, 1916, pp. 5–6.
16. Science-technology-society (STS) curricula were considered important reforms in the 1990s. Viewed as integrative and student-centered, presentation of STS in college and high school programs of study were common. John Yager documents STS as a reform in a book titled *Science/technology/society as reform in science education*; an introduction to the history of STS which is retrieved from http://www.sunypress.edu/pdf/53355.pdf
17. See image of Pres. Johnson and Ms. Loney at https://socialwelfare.library.vcu.edu/wp-content/uploads/2011/02/600px-ESEAJohnson-1.jpg

18. 1954 was a watershed moment in legal execution of equitable practices in education. In addition to the decision by the Supreme Court to render "separate" classrooms inherently unequal, organizations like the National Council for the Accreditation of Teacher Education (NCATE) were founded to provide leadership in helping develop teachers who would be prepared to uphold principles of equity, social justice and an ethic of care in the classroom. See NCATE Standards (2008) for more information at http://www.ncate.org/documents/standards/NCATE%20Standards%202008.pdf
19. A December 2017 search of the U.S. Congressional record online at https://www.congress.gov/ for "Elementary and Secondary Education Act of 1965" returns more than 4,000 results.
20. See the 2011 consensus study report published by the Committee on Underrepresented Groups and the Expansion of the Science and Engineering Workforce (U.S.), Committee on Science, Engineering, and Public Policy (U.S.), and National Research Council. *Expanding underrepresented minority participation: America's science and technology talent at the crossroads.*
21. Educational Longitudinal Study of 2002 available at https://nces.ed.gov/surveys/els2002/tables/APexams_05.asp
22. An important bill search feature on the Library of Congress website https://www.congress.gov/ is the ability to search related text for relevant legislation allowing researchers to explore archival information.
23. High School Longitudinal Study of 2009 Indicator 13: Advanced Placement and International Baccalaureate Coursetaking, Retrieved from https://nces.ed.gov/programs/raceindicators/indicator_rce.asp
24. National Academy of Sciences, National Academy of Engineering, & Institute of Medicine, 2007
25. New Jersey Plan for ESSA presentation materials are retrieved from http://www.state.nj.us/education/ESSA/plan/Overview.pdf
26. State profiles based on Common Core Data are available at https://nces.ed.gov/programs/stateprofiles/
27. New Jersey Administrative Code Title 6A: Chapter 7. Retrieved from http://www.state.nj.us/education/code/current/title6a/chap7.pdf
28. See "History of Abbott v. Burke" at http://www.edlawcenter.org/cases/abbott-v-burke/abbott-history.html
29. Gay, 1994.
30. Full bibliographic details are not provided however each work is widely available in a variety of open-source and peer-reviewed resources. Throughout this chapter, works are referenced in this way as an invitation to do additional reading.
31. Noted author and public intellectual James Baldwin delivered a speech to educators known as "A Talk to Teachers" in October 1963 that speaks to the needs of the "Negro child," framing education specifically to Negro children that opened discourse differently than in previous "poor children" discussions, making room for a critical race approach to education. Reprint is available through Yearbook of the National Society for the Study of Education, *107*(2): 15–20.
32. Banks, 1974.
33. Summary available at http://resources.css.edu/diversityservices/docs/levelsofintegrationofmulticulturalcontent.pdf

34. Full article available at https://education.uw.edu/sites/default/files/Review%20of%20 Research%20AERA.pdf
35. Bell, 1995.
36. Martin & Atwater, 1992.
37. Atwater, 1996.
38. Arámbula-Greenfield & Feldman, 1997.
39. Bailey, Scantlebury, & Johnson, 1999.
40. Bay, Staver, Bryan, & Hale, 1992; Kahle, 1998c.
41. Bell, 1995; Lynn, 1999.
42. Jacqueline Leonard, a mathematics educator, introduced an alternative to responsive/relevant/critical race in the canon of thought about CRP in her 2008 book *Culturally Specific Pedagogy in the Mathematics Classroom: Strategies for Teachers and Students*, Routledge. A second edition is currently in press. The notion of "specificity" implied in the title encompasses all of the former references.
43. The *Guide to Implementing the Alaska Cultural Standards for Educators* at https://education.alaska.gov/standards/pdf/cultural_standards.pdf
44. Stephens, 2001; at the time of this writing, Dr. Sidney Stephens' affiliation remains with the University of Alaska at Fairbanks although she is retired. Additional information about the legacy of her work is available online at http://mapteach.org/about/people/
45. Gummer & Champagne, 2006.
46. Settlage & Southerland, 2007.
47. Young, 2011.
48. Tate, 2001.
49. Mutegi, 2011; Rodriguez, 1997.
50. Gregory, 1922.
51. Available online at http://www.project2061.org/publications/bsl/online/index.php
52. Lee & Luykx, 2005. Although "non-mainstream" was used to describe students in 2005, current work from the first author reflects changes in descriptors to include "less privileged" and "non-dominant" when referring to the same populations of marginalized students. See https://www.nextgenscience.org/sites/default/files/PolicyNGSSforDiversityandEquity.pdf for additional insights.
53. Seminal works of Martin Haberman can be found through the Haberman Foundation website http://habermanfoundation.org/resources-media/articles/ (Retrieved December 2017).
54. "Enabling practices" as an alternative to "capacity building" generally associated with professional development.
55. See STEM Teaching Tools Practice Brief #43 (January 2017): "Why Do We Need to Teach Science in Elementary School," Retrieved from http://stemteachingtools.org/brief/43
56. Mutegi, 2011.
57. Shirazi, 2017.
58. Well defined by researchers like Rochelle Gutiérrez and Martin Haberman.
59. In a 2008 study exploring instructional time based on national data collected from a variety of districts, Jennifer McMurrer explains that since NCLB-era changes; 28% of the

elementary schools studied (based on 349 schools) decreased the amount of time dedicated to science instruction. Almost 66% of the schools reported a decrease of between 25 and 49 minutes per week when science instructional time was decreased AND mathematics/literacy instruction times were increased.

60. See Final Report prepared in 2005 of Mid-Continent Research for Education and Learning on High-Needs Schools retrieved through ERIC Database at https://files.eric.ed.gov/fulltext/ED486626.pdf
61. "Revenues, expenditures and poverty rate" data for large public school districts in 2015 are also Retrieved from https://nces.ed.gov/programs/digest/d15/tables/dt15_215.20.asp
62. Pringle, Brkich, Adams, West-Olatunii, & Archer-Banks, 2012.
63. Rational choice theory is an emerging theoretical framework for considering teacher actions in STEM. A recent study of teachers' choice to use technology as mediated by culture and emotions is available at https://doi.org/10.1007/s10639-015-9457-6
64. NARST Position on Equity in the Next Generation Science Standards https://narst.org/ngsspapers/equity.cfm
65. Tanner, 2013.
66. See "SF State Researchers Create a New Tool ..." at https://news.sfsu.edu/news-story/sf-state-researchers-create-new-tool-measures-active-learning-classrooms
67. The expansion of 5E learning cycle to 7E was proposed in 2003 by the National Science Teachers Association. See http://www.nsta.org/publications/news/story.aspx?id=48547
68. Saçkes, Trundle, Bell, & O'Connell, 2011.
69. Psychologist Albert Bandura has established that efficacy is developed through experience, mastery, verbal persuasion and psychological response. Each of these factors shapes beliefs about self. When these beliefs are substantiated by practices, realization of observable equity is possible, according to the Kahle metric.
70. I use the phrase "diversity texts" as a means to describe students' identity markers and/or social practices. Socially constructed markers like race, religion, and gender can be engineered by self or others to convey beliefs and understandings. Students' invented and conventional use of language and culture, styles of dress and communication/expression of self (or others) are indicators of social practice.
71. Barnes-Johnson, 2011.
72. Self-efficacy beliefs are typically measured by two sub-scales: personal beliefs about ability and outcome expectancy. The construct is well documented in a range of different fields.
73. Sazon! is a popular Mexican spice or condiment.
74. Boykin & Toms, 1985.
75. Barba, 1995, 1993.
76. Bennett, 1986.
77. Brown, 2017.
78. See the matrix created to show details about the 52 studies at http://z.umn.edu/1d5b
79. Artist Talk: Titus Kaphar at Princeton University (Recorded November 16, 2017) http://artmuseum.princeton.edu/video/artist-talk-titus-kaphar
80. Daugherty, 2013.
81. Twitter feed https://www.huffingtonpost.com/rob-furman/stem-needs-updated-to-str_b_5461814.html?ncid=engmodushpmg00000004

82. Blackley, Sheffield, Maynard, Koul, & Walker, 2017.
83. Decker, Mayer, & Glazerman, 2004; Hopkins & Heineke, 2013; Lewis & Gray, 2016; Mikulecky, Shkodriani & Wilner, 2004.
84. See "American Educational History: A Hypertext Timeline" by Edmund Sass (2011/2017) http://www.eds-resources.com/educationhistorytimeline.html
85. Walkley & Cox, 2013.
86. Problematizing discussions of gender normative and alternative dichotomies of social and physiological characterizations by scientists and science educators is necessary in 21st century identity politics. Basing STEM equity discussions primarily on gender when multiple cultural sub-strata exist within it is counterproductive and easily rejected as useful. Unpacking this idea is beyond the scope of this volume but deserves consideration. The National Science Teacher Association position statement on Gender Equity (http://www.nsta.org/about/positions/genderequity.aspx) has not been updated since 2003.
87. "Righting" is used here to describe formally structured political and informal stream-of-consciousness writing that addresses power structures and agency. When students are encouraged for example to participate in large authentic discourses that allow for the development of voice and change-agency, "righting" of "wrongs" is made more possible. An example of this type of enabling assignment is "Letters to the Next President 2.0" (https://letters2president.org/).
88. Berry, 2005.
89. Keolu Fox explains limits of genetic research from his perspective as a Native Hawai'ian and a lack of diversity in data (https://www.ted.com/talks/keolu_fox_why_genetic_research_must_be_more_diverse). Other discussions of Indigenomics consider the relationship of Indigenous peoples' perspectives on major value dissonances around sustainability, economics and human enterprise (http://www.rapidshift.net/1151-2/).
90. STEM for Global Leadership; retrieved December 2017 from https://www.ed.gov/stem
91. See https://www.whosampled.com/ for tools to compare sampled and original music.
92. ACESSE provides STEM Teaching tools as an open educational resource.
93. Ostrom, 1998, p. 2.
94. Fysh & Lucas, 1998.
95. Next Generation Science Standard HS-LS3-1: Heredity, Inheritance and Variation of Traits.
96. See "Genetic details of controversial three-parent baby revealed" by Sara Reardon (April 2017) Retrieved from https://www.nature.com/polopoly_fs/1.21761!/menu/main/topColumns/topLeftColumn/pdf/nature.2017.21761.pdf

References

21st Century STEM for Girls and Underrepresented Minorities Act of 2010, H.R. 6078, 111th Cong. (2010).

Arámbula-Greenfield, T., & Feldman, A. (1997). Improving science teaching for all students. *School Science and Mathematics, 97*(7), 377–387.

Atwater, M. (1996). Social constructivism: Infusion into the multicultural science education research agenda. *Journal of Research in Science Teaching, 33*(8), 821–837.

Bailey, B. L., Scantlebury, K. C., & Johnson, E. M. (1999). Encouraging the beginning of equitable science teaching practice: Collaboration is the key. *Journal of Science Teacher Education, 10*(3), 159–173.

Bang, M., Brown, B., Barton, A. C., Roseberry, A., & Warren, B. (2017). Toward more equitable learning in science: Expanding relationships among students, teachers, and science practices. In C. Schwarz, C. Passmore, & B. J. Reiser (Eds.), *Helping students make sense of the world using next generation science and engineering practices* (pp. 33–58). Arlington, VA: NSTA Press.

Banks, J. A. (2010). Foreword. In O. Lee & C. A. Buxton (Eds.), *Diversity and equity in science education: Research, policy and practice* (pp. ix–xiii). New York: Teachers College Press.

Banks, J. A. (1997). Multicultural education: Characteristics and goals. In J. A. Banks & C. A. M. Banks (Eds.), *Multicultural education: Issues and perspectives* (3rd ed., pp. 3–31). Boston, MA: Allyn & Bacon.

Banks, J. A. (1993). Multicultural education: Historical development, dimensions, and practice. *Review of Research in Education, 19*(1), 3–49.

Banks, J. A. (1989). Integrating the curriculum with ethnic content: Guidelines and approaches. In J. A. Banks & C. McGee-Banks (Eds.), Multicultural education: Approaches and guidelines (pp. 189–206). Boston, MA: Allyn & Bacon.

Banks, J. A. (1974). Multicultural education: In search of definitions and goals. Retrieved from https://files.eric.ed.gov/fulltext/ED100792.pdf

Barba, R. H. (1995). *Science in the multicultural classroom: A guide to teaching and learning.* Needham Heights, MA: Allyn & Bacon.

Barba, R. H. (1993). A study of culturally syntonic variables in the bilingual/bicultural science classroom. *Journal of Research in Science Teaching, 30*(9), 1053–1071.

Barnes-Johnson, J. M. (2011). *Efficacy-related beliefs and practices about equitable science teaching: A case study in an urban elementary school.* (Doctoral dissertation). Retrieved from Proquest Dissertations & Theses A & I (Order No. 3457862)

Bay, M., Staver, J. R., Bryan, T., & Hale, J. B. (1992). Science instruction for the mildly handicapped: Direct instruction versus discovery teaching. *Journal of Research in Science Teaching, 29*(6), 555–570.

Bell, D. A. (1995). Who's afraid of critical race theory? *University of Illinois Law Review, 1995*(4), 893–910.

Bennett, C. I. (1986). *Comprehensive multicultural education: Theory and practice.* Boston, MA: Allyn and Bacon.

Berry, J. W. (2005). Acculturation: Living successfully in two cultures. *International Journal of Intercultural Relations, 29*(6), 697–712.

Blackley, S., Sheffield, R., Maynard, N., Koul, R., & Walker, R. (2017). Makerspace and reflective practice: Advancing pre-service teachers in stem education. *Australian Journal of Teacher Education (Online), 42*(3), 22–37.

Boykin, A. W., & Toms, F. D. (1985). Black Child socialization. In H. P. McAdoo & J. L. McAdoo (Eds.), *Black Children: Social, educational and parental environments* (pp. 33–51). Beverly Hills, CA: Sage Publications.

Brabham, D. C. (2009). Crowdsourcing the public participation process for planning projects. *Planning Theory, 8*(3), 242–262.

Brown, J. C. (2017). A metasynthesis of the complementarity of culturally responsive and inquiry-based science education in K-12 settings: Implications for advancing equitable science teaching and learning. *Journal of Research in Science Teaching, 54*(9), 1143–1173.

Carello, J., & Butler, L. D. (2015). Practicing what we teach: Trauma-informed educational practice. *Journal of Teaching in Social Work, 35*(3), 262–278.

Daugherty, M. K. (2013). The prospect of an "A" in STEM education. *Journal of STEM Education: Innovations and Research, 14*(2), 10–15.

Decker, P. T., Mayer, D. P., & Glazerman, S. (2004). *Effects of teach for America on students: Findings from a national evaluation.* Institute for Research on Poverty Discussion Paper no. 1285–04. Retrieved from http://www.ssc.wisc.edu/irpweb/publications/dps/pdfs/dp128504.pdf

Dewey, J. (1916). *Democracy and education.* New York: Macmillan.

Fysh, R., & Lucas, K. B. (1998). Religious beliefs in science classrooms. *Research in Science Education, 28*(4), 399–427.

Gay, G. (2000). *Culturally responsive teaching: Theory, research, and practice.* New York: Teachers College Press.

Gay, G. (1994). *A synthesis of scholarship in multicultural education.* Urban Monograph Series. Retrieved from https://files.eric.ed.gov/fulltext/ED378287.pdf

Gregory, R. (1922). The teaching of science. *Science Magazine, 56*(1451), 433–439.

Gummer E., Champagne A. (2006) Classroom assessment of opportunity to learn science through inquiry. In W. W. Cobern (Series Ed.) & L.B Flick & N. G. Lederman (Vol. Eds.) *Science & technology education library: Vol. 25. Scientific inquiry and nature of science* (pp. 263–297). Dordrecht, the Netherlands: Springer.

Gutiérrez, R. (2002). Enabling the practice of mathematics teachers in context: Toward a new equity research Agenda. *Mathematical Thinking and Learning, 4*(2–3), 145–187.

Haberman, M. (2006). The special role of science teaching in schools serving diverse children in urban poverty. In W. W. Cobern (Series Ed.) & L.B Flick & N. G. Lederman (Vol. Eds.) *Science & technology education library: Vol. 25. Scientific inquiry and nature of science* (pp. 37–53). Dordrecht, the Netherlands: Springer.

Haberman, M. (2004). Can STAR teachers create learning communities? *Educational Leadership, 61*(8), 52–56.

Haberman, M. (1991). The pedagogy of poverty versus good teaching. *Phi Delta Kappan, 73*, 290–294.

Hopkins, M., & Heineke, A. J. (2013). Teach for America and English language learners: Shortcomings of the organization's training model. *Critical Education, 4*(12), 18–36. Retrieved from http://ojs.library.ubc.ca/index.php/criticaled/article/view/183908

Howe, J. (2006, June 2). *Crowdsourcing: A definition* [Weblog]. Retrieved from http://crowdsourcing.typepad.com/cs/2006/06/crowdsourcing_a.html

Kahle, J. K. (1998a). *Reaching equity in systemic reform: How do we assess progress and problems?* Research Monograph. ERIC Database: ED472341.

Kahle, J. K. (1998b). Equitable systemic reform in science and mathematics: Assessing progress. *Journal of Women and Minorities in Science and Engineering, 4*(2 & 3), 91–112.

Kahle, J. K. (1998c). *NISE brief: Measuring progress toward equity in science and mathematics education.* NISE Brief Vol. 2, No. 3. Madison, WI: National Institute on Science Education. Retrieved from http://www.wcer.wisc.edu/archive/nise/publications/Briefs/Vol_2_No_3/Vol.2,No3.pdf

Ladson-Billings, G. (1994). *The Dreamkeepers: Successful teachers of African American students.* San Francisco, CA: Jossey-Bass.

Landrum, T. J., Cook, B. J., Tankersley, M., & Fitzgerald, S. (2007). Teacher perceptions of the useability of intervention information from personal versus data-based sources. *Education and Treatment of Children, 30*(4), 27–42.

Lee, O., & Luykx, A. (2006). *Science education and student diversity: Synthesis and research agenda.* New York: Cambridge University Press.

Lee, O., & Luykx, A. (2005). Dilemmas in scaling up educational innovations with nonmainstream students in elementary school science. *American Educational Research Journal, 42*(3), 411–438.

Lewis, L., & Gray, L. (2016). *Programs and services for High School English Learners in Public School Districts: 2015–16 (NCES 2016–150).* U.S. Department of Education. Washington, DC: National Center for Education Statistics. Retrieved from http://nces.ed.gov/pubs2016/2016150.pdf

Lynn, M. (1999). Toward a critical race pedagogy: A research note. *Urban Education, 33*(5), 606–626.

Martin, H. J., & Atwater, M. M. (1992). *The stages of ethnicity of pre-service teachers and in-service personnel involved in multicultural education experiences.* ERIC Database: ED397203.

McMurrer, J. (2008). *NCLB year 5: Instructional time in elementary schools: A closer look at changes for specific subjects.* Retrieved from George Washington University, Center on Education Policy website https://www.cep-dc.org/displayDocument.cfm?DocumentID=309

Mikulecky, M., Shkodriani, G., & Wilner, A. (2004, December). *Growing trend to address the teacher shortage.* Denver, CO: Education Commission of the States. (ERIC Document Reproduction Service No. ED484845)

Mutegi, J. W. (2011). The inadequacies of "Science for All" and the necessity and nature of a socially transformative curriculum approach for African American science education. *Journal of Research in Science Teaching, 48*(3), 301–316.

National Academy of Sciences, National Academy of Engineering, and Institute of Medicine. (2011). *Expanding underrepresented minority participation: America's science and technology talent at the crossroads.* Washington, DC: The National Academies Press. Retrieved from https://doi.org/10.17226/12984

National Academy of Sciences, National Academy of Engineering, and Institute of Medicine. (2007). *Rising above the gathering storm: Energizing and employing America for a brighter economic future.* Washington, DC: The National Academies Press. https://doi.org/10.17226/11463

Ostrom, E. (1998). A behavioral approach to the rational choice of collective action: Presidential address, American Political Science Association. *The American Political Science Review, 92*(1), 1–22.

Pringle, R. M., Brkich, K. M., Adams, T. L., West Olatunii, C., & Archer Banks, D. A. (2012). Factors influencing elementary teachers' positioning of African American girls as science and mathematics learners. *School Science and Mathematics, 112*(4), 217–229.

Rodriguez, A. J. (1997). The dangerous discourse of invisibility: A critique of the National Research Council's National Science Education Standards. *Journal of Research in Science Teaching, 34*(1), 19–37.

Saçkes, M., Trundle, K. C., Bell, R. L., & O'Connell, A. A. (2011). The influence of early science experience in kindergarten on children's immediate and later science achievement: Evidence from the early childhood longitudinal study. *Journal of Research in Science Teaching, 48*(2), 217–235.

Settlage, J., & Southerland, S. A. (2007). *Teaching science to every child: Using culture as a starting point.* New York: Taylor & Francis.

Shirazi, S. (2017). Student experience of school science. *International Journal of Science Education, 39*(14), 1891–1912.

Stephens, S. (2001). *Handbook for culturally responsive science curriculum.* Fairbanks, AK: Alaska Science Consortium & Alaska Rural Systemic Initiative. Retrieved from http://www.ankn.uaf.edu/publications/handbook/handbook.pdf

Tanner, K. D. (2013). Structure matters: Twenty-one teaching strategies to promote student engagement and cultivate classroom equity. *CBE-Life Sciences Education, 12*(3), 322–331.

Tate, W. (2001). Science education as a civil right: Urban schools and opportunity-to-learn considerations. *Journal of Research in Science Teaching, 38*(9), 1015–1028.

Walkley, M., & Cox, T. L. (2013). Building trauma-informed schools and communities. *Children & Schools, 35*(2), 123–126.

Young, I. M. (2011). *Justice and the politics of difference.* Princeton, NJ: Princeton University Press.

CHAPTER TWO

Hip-Hop Pedagogy as a Framework to Support the Development of Science Geniuses

EDMUND S. ADJAPONG[1]

Abstract

This research[1] revolves around developing equitable pedagogical practices to provide urban youth, who traditionally have been marginalized in STEM disciplines, with access in STEM education. Research suggests that students from underrepresented ethnic groups traditionally fall behind their counterparts of less diverse backgrounds in major content areas, including science. In addition, urban students are less likely to be interested in the sciences partially because educators misunderstand the realities and experiences of urban students and as a result they are not able to demonstrate the relevance of science. This chapter suggests utilizing Hip-Hop/youth culture to develop effective and equitable pedagogical approaches to engage urban youth in STEM and to encourage urban students to view the field of STEM as attainable as they learn science through their culture.

1. Edmund S. Adjapong, Professor of Educational Studies
   Seton Hall University
   Edmund.Adjapong@shu.edu

Editorial Reflections

Research[2] has shown that when performance groups are the same as play groups, individuals thrive: learners whose formal and informal, curricular and extra-curricular, social and professional experiences are seamless do better academically. Boundaries between fun and learning disappear as motivations to succeed are tied to purposes that include social interaction, affirmation and acceptance. What then are the implications for individuals standing in social margins? Outside of social circles there are fewer points of entry into academic agency. Being invited and being included are not the same in STEM. I, Joy, started my first Ph.D. program at 21. I was on track to earn a Ph.D. in chemistry by the age of 25. In my second year, I switched institutions to a competitive program close to home. I was invited to enter the program with no teaching or research responsibilities. I was isolated. I failed. It took me nearly twenty years to recover from the shock and shame I felt.

Transitioning from the supportive STEM teaching and learning environment cultivated at an HBCU[3] to the different STEM teaching and learning environment I experienced at a PWI[4] was difficult. At the HBCU, I had been forced to be resourceful and develop strong ties with the other STEM majors. Small class size and mentor relationship was insurance for the students willing to persist. At the PWI, teaching assistants taught the classes with hundreds of other students. Professors' office hours were usually inconvenient and I knew no one. The participatory call-and-response style of questioning and review to which I had become accustomed was no longer available to me as an option when I needed help. The high-speed internet was not yet an everyday tool for learning. My previous strategies for learning were not working. I was trying to function outside of any cultural norm that I knew. The first time I raised my hand during a lecture, the stares and scowls I got from peers and instructors were unfamiliar—there was no care in play in this environment, I did not belong. I left that day, stopped coming to class and tried to learn on my own.

By the end of the semester, I decided to leave the program and find a job. Even though I had successfully completed enough coursework to earn a master's in chemistry, without supervisory guidance, I simply left. I eventually ended up at my former high school asking to be a lab technician. When I got there, I learned that I could be a teacher. I was "highly qualified" to be a secondary physical science teacher based on my grade point average and content degree. I took a test of content knowledge, took classes in the evening to develop my pedagogy and became a high school teacher. It came naturally to me in this environment. I was back on familiar terrain, with my former teachers who served as mentors for me, in my neighborhood/my community. I was being the kind of teacher I hadn't had in a long time; I was the cool teacher who dressed like my students and understood their fascination with Biggie-Tupac battles; in fact, I used pop culture as a basis for discussions for as many topics as I could. For my efforts, parents gave me a "Dream keepers Award."[5]

I, Joy, have come to think seriously about the borders we must cross when we exist "outside" of a circle,[6] *whether they be social, professional or academic. Border crossing is an ecological process, but capital-building requires us to move across multiple levels in horizontal and vertical directions.*

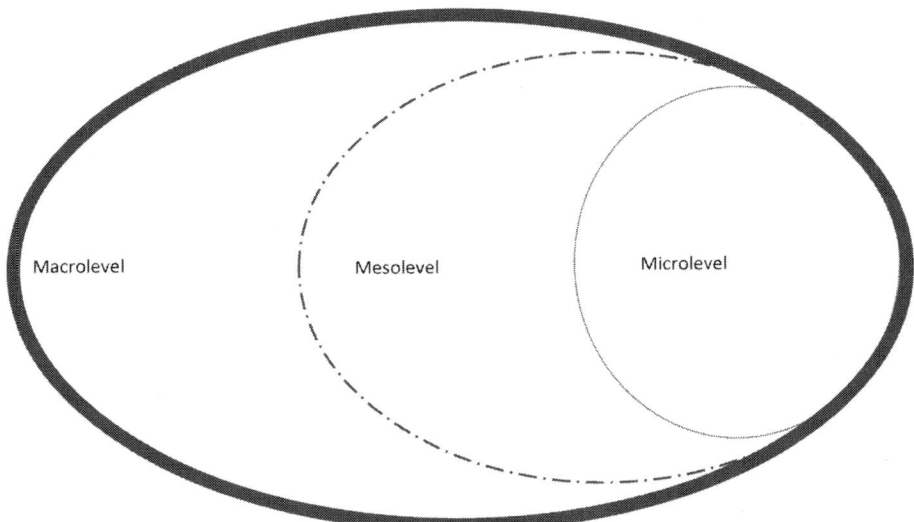

Figure 2.1: Macro/micro/meso levels.
Source: Barnes-Johnson

This organization of borders describes ecological levels of participation in STEM education programs. Characteristics include various macrolevel (macroscopic): impermeable barriers; mesolevel (intermediate) borders: perforated systems of constraints with clear pathways and flexible boundaries and microlevel (microscopic) borders: semipermeable barriers. Though described here for participation in STEM education programs, the same organizational ecologies exist for all types of participation and access.

Whether shown as vertical or horizontal, each ecological level represents movement/transcendence across levels of organization. Horizontally, movement is scaled to embody numbers of people and social comforts. Vertically, movement represents age and positioning in the context of youth[7] *and professional status—the separation of the world of play from the world of work. Examining my lived experience, at the micro-level I moved freely within small spheres of influence like my family, close affiliations and classrooms. At the meso-level, I was invited to participate broadly in larger communities, but found that not being included isolated me from larger (macro-level) activity. The borders themselves have been impenetrable barriers at times, sometimes thin semi-permeable veils, or perforated edges with a combination of gates and channels at other times. Teaching students how to cross these borders is a responsibility of adults in the STEM community at every level. Border crossing is a journey—field guides that use community relevant norms and practices within specific cultures help students build a strong sense of their own identity so they are able to cross the borders they face. Edmund Adjapong's chapter helps us think about a way to do just that.*

As a teacher educator, the first thing that jumps out at me, Janelle, is the fact that Edmund is teaching in the same community where he was raised, and he is a man of color. While many of us in the field are working to diversify the teaching population so that it more closely reflects the student population, there are many systemic forces that can be the impenetrable borders Joy describes. One way to work toward crossing these borders with STEM education is to foreground the voices of community-based educators. Edmund writes about building strong ties with his students through their shared cultural practices, and utilizing those relationships to facilitate the students' access to STEM learning. Statistically speaking, however, this is not possible for most current teachers—we are disproportionately White, and our classrooms are becoming more and more non-White. So for those of us who are not members of the communities where we teach, how can we build trust with our students in order to strengthen our ties, and better facilitate the learning they need to cross borders?

When I, Janelle, moved back from my first teaching job in Guatemala City, it was late autumn 1999. During my teacher preparation program, we had visited some of the schools in an inner city Phoenix district, and that is where I wanted to teach. Unaware of it at the time, looking back I realize I must have been subconsciously inspired by the romantic notion of "saving those poor kids of color" that somehow continue to be remade into a "new" movie every couple years. A new school had opened in the district, and some of the teachers from this school had taken positions there. I was offered one of the vacant spots at the middle school, and I started mid-year. The students, who were about 80% Latinx and 20% African American, called me out on my outsider status instantly. "Miss, where are you from?" and "You're not from around here!" I was not from their neighborhood, or anywhere close for that matter, and I didn't try to tell them otherwise.

I had assumed that since I had successfully moved to Central America and back, that taking a job in a different neighborhood of my own city would be fairly seamless. To be honest, I experienced more culture shock at the school than I'd had moving abroad. My best friend growing up was Black, as was my college roommate. I'd lived in a Spanish speaking country. None of that mattered. I did not get a pass from these kids. I couldn't just explain to them that they should trust me. And quite frankly, if I had tried to implement Hip-Hop pedagogy, they would have died laughing. It would not have been authentic in the least for me as their teacher.

What I did do was be patient, and recognize that I was the outsider in their community. I realized that these students had no reason to trust me, and plenty of reason not to. I treated them with respect, and I worked to push their learning to show them that I cared about them with high expectations. For some of the non-English speaking students, that meant providing free English classes after school. For the students who played sports, it meant going to as many games as I could, cheering them on and trying to connect with their families when possible. And for the student who showed me the toughest exterior, it meant helping him to read his native language. Jorge was an eighth grader, and one of several students at the school who was on probation. He was vocal about his dislike of school, and he especially hated reading. He spoke Spanish at home and with his friends, but like many of the bilingual students, had only been taught to read in English. I wondered if providing him with appropriate Spanish texts (that didn't feel babyish) would catch his interest. They did. He started reading more and more books in Spanish, and then even started doing more reading in English. I strengthened my ties to my students over time. There were certainly students who I was never able to make a connection with but all the students saw that I was being my authentic self, while respecting their identities, and getting to know them on a personal level. Wherever the teacher is from, being responsive to students' needs is a necessary component of pedagogy.

Adjapong's chapter is not a recipe for what to do but a model of how to learn from the community where you are. Claiming expertise will probably not help outsiders gain entry; listen, and be a learner. Work to understand the language and codes of the community, build relationships, explore the open doors when available and respect the veils when they are in place.

Introduction

As a science educator teaching in the same urban community where I grew up and attended public school, I notice that my students have a very similar connection to Hip-Hop culture as I had as an adolescent. As a child who was born and raised in the Bronx, NY, the birthplace of Hip-Hop, I noticed and participated in all forms of Hip-Hop expression in my community, but outside of schools. It is important that I highlight that I engaged in Hip-Hop forms of expression in my community, but outside of schools because my Hip-Hop identity was not welcomed in school. As a child, I would walk up the block to my elementary school with a group of friends rapping the latest Hip-Hop song, but when we arrived at school we were always told to stop and that rapping in the hallways was unacceptable.

Growing up, Hip-Hop was everywhere in my urban community. I lived directly across the street from the largest park in my neighborhood. During the summer months, I would hear older "throwback" Hip-Hop songs booming from the large speakers from the park into my apartment window. Cars in my neighborhood drove by blasting the latest and most popular Hip-Hop songs so loudly you could hear, and sometimes see the car rattling. When playing basketball in the park across the street from my apartment complex, there was always a chance that a cypher could begin at any moment. Everyone dressed to reflect the most popular Hip-Hop clothing trends—fresh all white Nike "Uptown" sneakers and crisp white tees were imperative for every young man's summer wardrobe.

As a science educator, who teaches in the same community where I was raised, I notice that my students share similar connections to Hip-Hop culture as I. Paradoxically, as a science educator I also notice student disengagement in schools, more particularly in STEM classrooms. In recognizing students' disengagement in STEM classrooms, it is difficult to ignore the rates of science achievement as it relates to students of color, who predominantly populate the urban communities where they attend school. Educators are aware of the persistent achievement gaps or as Gloria Ladson-Billings (2006) refers to "educational debts" that exist between non-dominant students and their counterparts of less diverse social settings, especially if we consider students' performance on standardized exams as the chief marker of achievement. According to the National Science Foundation (2015), in the fourth grade there is a 14% gap of achievement in science between Black students as compared to their White counterparts. The gap of achievement in science persists as it increases to 17.4% when comparing Black and White students in the eighth grade. The National Science Foundation reported a similar trend when considering the achievement in science between Latinx students and White students, reporting a gap of achievement of 12.5% in the fourth grade and a gap of 13.9% in the 8th grade. When identifying gaps of achievement among groups of students who have historically been socially marginalized, it's particularly important to recognize the root of the achievement gaps.

Educational Debt in Science Education

In her 2006, presidential address to the American Educational Research Association, Gloria Ladson-Billings (2006) expressed the need for educators and policy makers to reframe their thoughts around the idea of the achievement gap, which the National Governors' Association (2005) describes as "a matter of race and class" that persists between non-dominant and disadvantaged students and their White counterparts. In drawing sharp distinctions to the status quo,

Ladson-Billings believes we should refer to the existing achievement gap as an education debt. She argues that when educators discuss and propose remedies to improve the achievement gap, which mainly focus on the disparities of students' standardized test scores; we are solely focusing on student achievement without questioning the systems and structures in place that may account for disparities in achievement measures. Therefore, if we only interrogate the achievement gap as a phenomenon birthed from non-dominant and disadvantaged students performing poorly when compared to their White counterparts, the burden of poor academic achievement lies solely on the student, without considering the systems, structures and spaces which provide urban students a so-called valuable education that is ineffective. Ladson-Billings (2006) suggests, "looking at the [achievement] gap is a misleading exercise." Studies show that urban schools, which mainly serve students from underrepresented ethnic groups, have fewer experienced teachers, receive less resources and teachers have less control over the curriculum, as compared to schools in suburban communities.[8] By focusing on improving the education debt of urban schools we can ultimately improve the academic achievement of students who attend urban schools, especially as it relates to science.

Once it is understood that there is a significant education debt between non-dominant students and their White counterparts writ large, we can address the fact that this education debt persists in specific content areas such as science. Researchers find that schools that serve mainly non-dominant groups of students offer fewer science courses, and therefore offer less opportunity for students to succeed and experience science.[9] Moreover, Norman et al. (2001) argue that the science education debt for students of color is a function of marginalized groups' disadvantaged position in the United States' society. Groups of people who have a disadvantaged position in society tend to suffer socioeconomic hardship, and the stigma of inferiority; "In urban settings schools in impoverished neighborhoods underperform relative to schools in more affluent settings."[10] A low number of students of color successfully complete high school and go on to pursue a science related degree in college. Therefore, only about 17% of scientists and engineers in the United States are from non-dominant communities.[11]

Undoubtedly, there is a crisis as it relates to the number of students of color who are persisting and pursuing careers in STEM, especially when compared to their White counterparts. I argue that while there is a crisis, as stakeholders in education, we must look at this crisis considering the systemic abuse that has negatively affected students of color, who have been pushed to margins of the fields of STEM. Rather than focusing on standardized exams we must focus on instruction and pedagogy, the way that we are teaching STEM to students of color and students of the 21st century.

Previous White House Focus on STEM

Due to an increasingly competitive global economy, growing challenges in sustainability, and the effects of the education debt in science education, there has been periods of specific emphasis on improving outcomes in science, technology, engineering and mathematics in recent years. In 2015, President Obama announced over $240 million to prepare students to excel in the STEM fields, especially those from underrepresented ethnic groups.[12] The United States Department of Education reports that the United States is falling behind internationally in both mathematics and science, which is unacceptable considering the global economy. Reports show STEM occupations are projected to grow by 17% from 2008 to 2018, compared to 9.8% growth for non-STEM occupations.[13] Concurrently, African American and Latinx students' expressed interest in STEM has decreased significantly. It is now lower compared to ethnic counterparts and is expected to remain lower during the coming years.[14] Therefore, due to students' lack of interest in STEM, especially by students of color, it is highly likely that they will not acquire the skills needed to compete for future STEM jobs and ultimately there will be fewer jobs available for them. In order for urban students to successfully compete in the competitive global economy, science educators must utilize innovative approaches to teach and engage students, especially non-dominant students, in science.

Equally as important as engaging and preparing students of all backgrounds to pursue careers in STEM is the idea of developing a more science minded society. It is increasingly important that we encourage all students, regardless of their pursuit of STEM careers, to be more science literate and develop the traits of the average scientist, which include imagination, persistence, curiosity, ethics, intelligence, and perseverance. We have daily experiences that science can be used to explain and make sense of, which is why it is important to encourage all students to be more scientifically literate.

Hip-Hop Pedagogy

When considering teaching strategies and approaches to engage youth in STEM, we are at an intersection in where we must recognize that traditional teaching approaches that have been used to engage students have failed in engaging students in STEM. I argue that in order to increase engagement in STEM, and therefore encourage more students to pursue careers in STEM, we incorporate students' culture in the teaching and learning. Research demonstrates that students have preconceived stereotypes of scientists that they cannot relate to. "Multiple lines of evidence suggest that the attitudes students develop regarding science and scientific careers are important factors in predicting persistence and success in science, technology, engineering, and math."[15] In addition to not identifying as scientists, traditionally students have not been engaged in science practices. Not having a command on STEM learning methods (or being taught in a style) to which they can directly relate, students simply don't participate.

In understanding the current landscape of science education, I present Hip-Hop pedagogy as a way of teaching utilizing Hip-Hop, which is youth culture. I define Hip-Hop pedagogy as a culturally relevant way of authentically and practically incorporating the creative elements of Hip-Hop into teaching, and inviting students to have a connection with the content while meeting them on their cultural turf by teaching to and through their realities and experiences. Hip-Hop pedagogy is grounded in the framework of reality pedagogy[16] and culturally relevant pedagogy.[17] In this innovative approach to teaching and learning, I identify pedagogical approaches that are directly linked to the creative elements Hip-Hop culture (MCing, graffiti art, breakdancing, DJing and knowledge of self) and practices in which urban youth engage in daily outside of school walls. From culturally relevant pedagogy, Hip-Hop pedagogy draws a focus on understanding Hip-Hop & youth culture. Culturally relevant pedagogy encourages teachers to immerse themselves so deeply in the culture of the specific students through actual engagement with the students, that it becomes second nature to find ways to develop students' interest in, and natural affinity for, science. From reality pedagogy, Hip-Hop pedagogy draws a focus on the teacher learning about the authentic realities of students and teaching utilizing students' authentic realities and culture to better engage them in science.

Although Hip-Hop is the most consumed genre of music by listeners across the world, many individuals identify it as only a musical genre.[18] Hip-Hop is a culture that has been in existence for over 40 years. Hip-Hop was birthed in the midst of social and economic crisis in the Bronx and was an effort to build strong communal ties in times of economic and social hardship, urban youth attended block parties to escape their unfortunate realities. Since its conception, Hip-Hop has impacted and empowered youth populations across the globe, especially youth of marginalized groups.[19] Hip-Hop pedagogy provides practical and tangible tools that STEM educators can use to teach students of the Hip-Hop generation.

Ultimately, if teachers are invested in using Hip-Hop pedagogical practices that are directly anchored in the realities of students it allows for the formation of "weak ties" and "strong ties" between teachers and students. There are links between individuals and groups within every social network that are categorized as "strong ties" or "weak ties."[20] "Strong ties" correspond to the connections individuals or groups who are "friends" have in common. While, weak ties correspond to "acquaintances" who do not have much in common that would normally connect them.[21] Students of the Hip-Hop generation who participate in the same culture (Hip-Hop) would be known to have "strong ties" because they share the same cultural practices forming dense, close knit networks that facilitate trust and cooperative exchanges. Teachers who do not identify as part of the Hip-Hop generation, while teaching student who do identify as part of the Hip-Hop generation would be known to have "weak ties" with their students because the teacher does not share the same cultural practices with students, but the teacher engages with students on daily basis, which encourages a "weak tie." When teachers use teaching approaches that are anchored in the culture and realities of their students it provides an opportunity for the teacher to gain insight on the students cultural practices and convert the "weak ties" that already exist largely because of generational differences into "strong ties" overtime. The creation of these dense networks between students and the science educator can allow for a positive exchange of science content and provide a space for educators to authentically learn about the realities of their students. As Hip-Hop pedagogy is anchored in the culture and practices of Hip-Hop, outlined in this section are practical approaches for each of the five creative elements (MCing, graffiti art, breakdancing, DJing and knowledge of self) Hip-Hop.

Element 1: The MC (Master of the Classroom and Master of Content)

In Hip-Hop the MC is known as the Master of Ceremonies. The MC is the person, often the rapper, who engages the crowd with their lyrical content. All MCs approach the role of MCing differently using their unique style. Oftentimes, when an MC is performing to an audience, they are accompanied by a fellow MC whose essential role is to provide support and to be a professional in terms of knowing and understanding the musical content; these roles showcase meaningful performance for the audience. While considering Hip-Hop pedagogy, we look at the MC as the Master of the Classroom and the Master of Content.

Co-Teaching

Co-teaching is a teaching approach, most commonly used in education and is defined as "two or more professionals delivering substantive instruction to a group of students with diverse learning needs."[22] The common goal for two co-teachers is to allow the responsibilities for instruction to be shared between the two professionals. When utilizing Hip-Hop pedagogy as an approach to teaching, the student is identified as a professional and expert in the science classroom. The responsibilities for instruction are shared between both the teacher who is normally viewed as the main authority figure of the classroom and a student. The goal is to empower the student to be a Master of Content and therefore a Master of the Classroom and provide them with the opportunity to share science content to their peers. In doing this, the student feels a sense of empowerment and excitement that will allow them to take an increased responsibility for their own learning and participation to enhance their science content knowledge.[23] Co-teaching with a student, as opposed with another teacher, increases instructional options, provides students with the opportunity showcase their mastery of the content as they support their peers to gain that same mastery of content. In addition, this kind of co-teaching in itself is a culturally relevant approach in the sense that the student who is now deemed the professional is a part of the same population that is both receiving and delivering the instruction, a mechanism of co-authorship and accountability. As we provide students with the opportunity to demonstrate their brilliance[24] in science by teaching their peers, the teacher is also provided an opportunity to learn from the students' in the analogies and methods that they use to engage their peers. In Hip-Hop pedagogy, co-teaching was supported using the following steps:

Before class:

- A student who volunteered to be a co-teacher is given a lesson plan to review for homework in preparation to teach the class the following day.
- The teacher performed a quick review of the lesson plan with the co-teacher to ensure that content is reflected accurately.
- The student was responsible for enhancing that lesson plan so that it can reflect their "teaching style."

During class:

- The teacher sits in a student's seat in a place that is prominent in the classroom and in the view of the co-teacher.
- The teacher pays close attention to parts of the lesson where the content delivered and guides the instruction (by raising a hand as a traditional student would) only when there are issues with the content (Emdin, 2011).

Call-and-response

In Hip-Hop, to engage the audience, traditionally the MC will use call-and-response during their performance as a way for audience members to participants during the performance. This exchange between the MC and the audience generates high energy and allows every audience member to participate in the exchange, that has been coined positive effervescence.[25] This state of energy bears witness to the "spontaneous verbal and non-verbal interaction between speaker and listener in which all of the statements ('calls') are punctuated by expressions ('responses') from the listener."[26] Responses from the audience can follow from a speaker specifically requesting them, or they can be unsolicited and spontaneously interjected into the ongoing interaction.[27] Call-and-response is a popular approach and is commonly used in Hip-Hop. Several studies show call-and-response to be effective in teaching students, especially in urban communities.[28] Call-and-response is considered integral to communicative behavior and functions as an expression of identity and as a means of conveying cognitive information among those who identify as the Hip-Hop generation.[29] When utilizing Hip-Hop pedagogy, call-and-response can be used as a classroom management tool, to generate positive emotional energy among students and as an approach to reinforce content knowledge. As a classroom management tool, call-and-response can be used to gain the attention of the entire class. To reinforce content knowledge, call-and-response can be used to review and reinforce content knowledge of simple science definitions. It's important to enact call-and-response at least three times when reinforcing content knowledge to allow students who may not know the specific call-and-response and opportunity to participate.

> *Call-and response to reinforce content knowledge:* Newton's first law of motion
> Teacher: An object that is in motion
> Students (in unison): Stays in motion
> Teacher: An object that is at rest
> Students (in unison): Stays at rest unless acted upon by an unbalanced force
> *Repeat two more times.*
>
> *Classroom management:* To gain the attention of students when necessary.
> Teacher: If you can hear my voice clap once
> Students (in unison): [Clap] Teacher: If you can hear my voice clap twice
> Students (in unison): [Clap] [Clap]
> Teacher: No music
> Students (in unison): [Clap]…[Clap] [Clap]…[Clap]

The clapping rhythm used in this call-and-response pattern originated from a classic Hip-Hop dance song entitled "No Music" by a Harlem rapper named Voice of Harlem.

Element 2: Breakdancing (Breaking)

The performance aspect of Hip-Hop is pivotal to the culture. During the conception of Hip-Hop, breakdancing crews were formed where the best breakdancers within each group would battle dance wise.[30] Breakdancers would dance to the rhythm of the beat played the neighborhood Disc Jockey (DJ) or from a loud boom box at anytime and anyplace. Breakdancers took the art of dancing seriously and always strived to perfect their moves. They danced faster, developed more complex moves and improved their form.[31] Although breakdancing is not currently as popular as it once was in the 1970s and 1980s, I argue that this style of dancing has evolved into contemporary Hip-Hop dance, which continues to be a pivotal part of Hip-Hop culture. The intricate, well thought out, well performed dances that Hip-Hop dancers perform demonstrates a kinesthetic aspect of Hip-Hop culture.

There are four stages of cognitive development which Bruner (1966) and Piaget (1951) describe as ways in which humans assimilate knowledge about their surrounding environment—through four sensory modalities: visual, auditory, visual/iconic and kinesthetic. Kinesthetic learners prefer "learning achieved through the use of experience and practice. In other words, the kinesthetic learner has to feel or live the experience in order to learn it."[32] Kinesthetic learning involves the physical manipulation of objects or the body, like a breakdancer.[33] Through breakdancing and contemporary Hip-Hop dance, urban youth learn how to manipulate their body. In doing so, youth who follow Hip-Hop culture, naturally communicate well through body language and [can] be taught through physical activity, hands-on learning, acting out, and role playing.[34]

When utilizing Hip-Hop pedagogy, students engage in kinesthetic learning activities that provide them the opportunity to physically manipulate objects and their body alike, similar to Hip-Hop breakdancers, with the goal of better understanding and engaging in science content. For example, the concept of states of matter can difficult for students to understand as we are unable to see molecules and how they interact. To provide students with an opportunity to engage in a kinesthetic activity to gain a better understanding of states of matter we should encourage students to act as individual molecules. Once students take on the persona of a molecule have the students as a class act as a solid. When students are acting as a solid, all students should be in close proximity, shoulder to shoulder and vibrating in place acting exactly like solid molecules. When students are acting as liquids, they should all be in the same area, assuming a fixed shape, but moving. Finally, when students are acting as a gas, they should be running around classrooms at high speeds and bouncing off of one another, exactly like gas molecules. These experiences allow students to engage in and gain a better understanding of science content, while engaging in a kinesthetic activity, just as Hip-Hop dancers.

Element 3: Graffiti Art

Graffiti art is an aspect of Hip-Hop culture that has not been has not been nearly as received as the other creative elements. The art of graffiti became popularized in New York City by the late 1960s where urban youth participated in tagging their street alias on the walls of urban neighborhoods, train cars, and storefronts because they enjoyed the attention their art received perhaps making them feel like local celebrities.[35] Graffiti art provides urban youth an opportunity to be expressive in their communities. Cultural scholars like Jeff Chang (2005) often suggest that graffiti artists may find it liberating to climb tall gates and slip under fences to create murals that represent them and tell their story visually. Identified as *reverse colonization*, graffiti artists and cultural scholars like Greg Tate[36] often talk about how mural art and music help marginalized people reclaim their communities when they have been taken away; a key example exists in New York City's planning efforts to build the Cross Bronx Expressway in the Bronx. Led by Robert Moses[37] in the mid-1900s, tens of thousands of Bronx residents were displaced by the Cross Bronx Expressway project.

In recent years, educators have been focusing on incorporating the arts into the STEM acronym, changing it to STEAM (Science, Technology, Engineering, Arts, Mathematics). Educators[38] suggest that "art and science are intrinsically linked" and students are able to better their understanding of science content through creating their own artistic representations of the science content. In support of incorporating art in the teaching and learning of science, science educators note that the "visual arts just seems to really be able to home in on the concept, taking it from the abstract to the concrete, so students are really able to understand it."[39] When utilizing Hip-Hop pedagogy, students are charged with tasks where they engaged in the visual arts, similar to graffiti artist, to work through and demonstrate their understanding of science content. Using visual art as a pedagogical approach to support students understanding of science content allows students to "make representations to: express their thoughts, feelings and perceptions; show relationships and changes; and make explanations and predictions."[40] An example of a graffiti art piece in the science classroom would be to have students draw their representation of the law of conservation of energy. The law of conservation of energy states that energy can neither be created nor destroyed; rather, it transforms from one form to another. Students can be encouraged to draw a visual representation of energy being transferred from one form to another. An example of this would be electrical energy from an power outlet being transferred to thermal energy from an iron. Students are able to gain a deeper understanding of science content when they are tasked to draw their own representation of the content and make explicit connections to world around them.

Element 4: DJ (Disc Jockey)

At its core, the DJ is responsible for supporting other creative elements including the Master of Ceremonies (MC) and the b-boys/b-girls. The disc jockey (DJ) is arguably the most important creative element of Hip-Hop culture. The DJs primary responsibility is playing and controlling the music, the rhythm, and the beat to which the MC adds their lyrical content to produce a completed product. The DJ is also responsible for finding the break in the beat, the moment in the song where only the drums are present, to provide an optimal rhythm for the b-boys/b-girls to showcase their best dance moves. Most importantly, the DJ is responsible for gauging the mood of a crowd and playing the perfect arrangement of songs to harness the crowd's energy. Jeff Chang's (2005) depiction of the conception of Hip-Hop describes DJ Kool Herc,

> Like any proud DJ, he wanted to stamp his personality onto his playlist. But this was the Bronx. They wanted the breaks. So, like any good DJ, he gave the people what they wanted, and dropped some soul and funk bombs. [People] were packing the room. There was a new energy. (p. 85)

Chang describes DJ Kool Herc as a DJ who incorporated his personality into his playlist while playing songs to the crowds' preference. Chang also explains that DJ Kool Herc was responsible for harnessing a new energy that was attractive to a crowd of people. When utilizing Hip-Hop pedagogy, students are to be in charge of being the DJ of the classroom and harnessing energy among their peers by creating playlist that are played during class. Teachers allow students to curate a class playlist of their favorite music instrumentals (music without lyrics). This can be done by creating a survey for each class and asking students to list their favorite songs, of any genre. A playlist of instrumentals can be created using YouTube for each class that is taught. The student-curated playlist should be played during class as background music when students are completing individual and group task. The goal of utilizing a playlist curated by students is to harness the same form of energy as a traditional DJ within the classroom. Also, a student-curated playlist provides an opportunity for teachers to gain knowledge about students' interests that they would not have learned otherwise.

Element 5: Knowledge of Self

Knowledge of self is the last creative element of Hip-Hop culture. Afrika Bambaataa is a DJ and is known as the grandfather of Hip-Hop (Kugelberg, 2007).[41] Bambaataa is also known for founding the group called the Zulu Nation in the early 1970s, which was composed mainly of a group of socially and politically aware rappers. Bambaataa and the Zulu Nation have defined knowledge of self as a central component of Hip-Hop culture. Bambaataa along with many Hip-Hop purists believe that knowledge of self is central because participants of Hip-Hop culture must remember that Hip-Hop was created based on authenticity and as a social political movement. Rap music represents a small percentage of Hip-Hop culture, which is known to have been commercialized and therefore slightly removed from being nested in the authenticity of Hip-Hop culture. Essentially, knowledge of self is central to Hip-Hop as it encourages participants of Hip-Hop culture to be aware of who they are, be authentic to themselves and be confident in oneself to make social political change for the community.

While it is reported that urban students' interest in STEM continue to decrease,[42] it is important that urban students increase their engagement in STEM and begin viewing the field of STEM as visible and accessible. Through the implementation of Hip-Hop pedagogy in a urban science classroom, I suggest that students will increase their engagement as it relates to STEM and increase their knowledge of scientific processes in the science classroom, which will allow them to be more comfortable navigating STEM spaces outside of the science classroom. Students would be confident in their skills and abilities as related to STEM and be confident enough to pursue a career in STEM if they chose to, rather than not wanting to pursue a career in STEM because they are not engaged.

The Development of Science Geniuses

As a Hip-Hop based intervention, The Science Genius BATTLES (Bring Attention to Transforming Teaching, Learning and Engagement in Science) program is anchored in Hip-Hop pedagogy. The Science Genius Program serves as an in school intervention to engage youth who are traditionally disengaged in science through Hip-Hop, especially rap. The program involves students writing and performing science-themed raps and participating in a final competition to showcase their science content knowledge.

The Science Genius program is an international program with participating sites including New York City, St. Lucia, Toronto, Calgary and Jamaica. The obstacles around the lack of engagement in science is a phenomenon that is found in all spaces where students who have be marginalized by social structures. Students who participate in the Science Genius Program represent all ethnic backgrounds and mainly identify as part of the Hip-Hop generation and listen to, write or engage in Hip-Hop music or one of its creative elements.

The Science Genius Program is implemented in addition to the traditional science curriculum. Science educators are provided professional development and are trained on engaging their students in this Hip-Hop based intervention. Teachers served as science content experts to students as they wrote their science-themed raps. Over the course of the semester students have created various science-themed raps that reflect their understanding of the multiple concepts that were taught in their science class. Students work collaboratively to check each other's science-themed raps for accurate science content and offer their peers constructive feedback in improving their science-themed raps. This is similar to a peer-review process that traditional scientists would engage in to publish their findings in research. This allows students to engage in the metacognitive, higher order thinking, process as they are evaluating their peers science-themed raps.

Toward the end of the semester, students are allowed to perform their science-themed raps in front of a large audience that is composed of students, youth from the community, teachers, family and community members. Students' science-themed raps are then judged by notable community members and Hip-Hop artists for science content and performance. Students shared that they feel like celebrities when performing their raps in front of such a supportive crowd, but also like Science Geniuses, as they are demonstrating their science content knowledge not only to their teachers, but to their entire community. Through research[43] we found that students make explicit connections between their realities and science content demonstrating ownership of learning and making connections. We also found that students were provided an outlet to disclose emotion through their science-themed raps. Students used emotion and science content to reframe their science identity/perceptions of science. Students used their rap and performance to push against the traditional idea of scientists and identify as individuals who are well-versed in both Hip-Hop and science.

Ultimately, the Science Genius program provides an innovative approach to teaching and learning, while affording students an alternative method for becoming engaged in the sciences. Science Genius utilized Hip-Hop pedagogy as it creatively used the elements of Hip-Hop as a guide to engage students in science. Students perform their raps, they dance, they become the MC (Master of Ceremonies) in the Hip-Hop sense and the MC^2 (Master of Content and Classroom) as Hip-Hop pedagogy is framed. The Science Genius Program serves as an exemplar of how we can utilize students' culture within the curriculum to teach science.

Discussion Questions

- Describe teacher and student actions for paradigms of culturally relevant pedagogy and Hip-Hop pedagogy as presented in the article.
- Reality pedagogy provides a context for the reclamation movement. How do each provide agency in STEM?

- What is the role of adult stakeholders in moving socially or academically marginalized students or groups into STEM discourse using the elements of Hip-Hop?
- Many teacher evaluation standards expect teachers to find ways to know their students. What strategies could you use to get to know your students that incorporate their primary ways of communication and self-expression?

Notes

1. Adjapong, 2017.
2. Chen, Chang, & He, 2003.
3. HBCU: Historically Black College or University
4. PWI: Predominantly White Institutions
5. Ladson-Billings, 1994.
6. Cipher (cypher) is a reference used to describe a secret or protected code or zeroes. In Hip-Hop/R&B, the term *cipher* has multiple meanings that allude to circular shape or battle zones where dancers for example are centered for a small audience.
7. Lamar, 2015 provides a relevant example of these levels. Kendrick Lamar transcends time and message by integrating a 1994 interview of Tupac Shakur into a current track.
8. Lippman, Burns & McArthur, 1996; Freedman & Appleman, 2009.
9. Norman, Ault, Bentz, & Meskimen, 2001.
10. Norman et. al., 2001, p. 105.
11. National Science Foundation, 2015.
12. Fact Sheet, 2015.
13. Langdon, McKittrick, Beede, Khan & Doms, 2011.
14. Munce & Fraser, 2012.
15. Schinske, 2015.
16. Emdin, 2016; Emdin, 2010.
17. Ladson-Billings, 1994.
18. Hooton, 2015.
19. Adjapong & Emdin, 2015; Dunley, 2000.
20. Coleman, 1988.
21. Easley & Kleinberg, 2010.
22. Cook & Friend, 1995, p. 25.
23. Lave & Wenger, 1991; John-Steiner & Mahn, 1996.
24. See also *The Brilliance of Black Children in Mathematics* (2013) by Editors Jacqueline Leonard and Danny B. Martin.
25. Bourdieu, 1986.
26. Smitherman, 1977; p. 104
27. Foster, 1989.
28. Foster, 2002; Piestrup, 1973.
29. Cazden, 1988.

30. To consider individuals "dance wise" is to accept the capital of rhythmic and kinesthetic intelligence.
31. Fresh, 1984.
32. Murphy, Gray, Straja, & Bogert, 2004.
33. Gardner, 1993.
34. Lane, 2008.
35. Chang, 2005.
36. Tate & Goodison, 2012.
37. Campanella, 2017.
38. Alberts, 2008.
39. Robelen, 2011.
40. Nelson & Chandler, 1999, p. 41.
41. Johan Kugelberg collated photographs and artifacts in the 1970s as the Hip-Hop movement was being born, in the same way as researchers might consider ethnography. This volume is described as a curated collection of the voice and work of Joe Conzo, Afrika Bambaattaa and others as they birthed a movement. Attributions of this volume have been made to the various artists of that time and citations for this text may vary.
42. Munce & Fraser, 2012.
43. Emdin, Adjapong & Levy, 2016.

References

Adjapong, E. S. (2017). *Bridging theory and practice: Using Hip-Hop pedagogy as a culturally relevant approach in the urban science classroom.* (Unpublished doctoral dissertation). Columbia University, New York.

Adjapong, E. S., & Emdin, C. (2015). Rethinking pedagogy in urban spaces: Implementing Hip-Hop pedagogy in the urban science classroom. *Journal of Urban Learning Teaching and Research*, 66(11), 66–76.

Alberts, R. (2008). Discovering Science Through Art-Based Activities—Earth's Changing Surface—Beyond Penguins and Polar Bears. Retrieved March 19, 2016, from http://beyondpenguins.ehe.osu.edu/issue/earths-changing-surface/discovering-science-through-art-based-activitieshttp://beyondpenguins.ehe.osu.edu/issue/earths-changing-surface/discovering-science-through-art-based-activities

Bourdieu, P. (1986). The forms of capital Handbook of theory and research for the sociology of education (pp. 241–258).

Bronx Museum of the Arts, Walker Art Center, & Spelman College. Museum of Fine Art. (2001). *One planet under a groove: hip hop and contemporary art.* Bronx Museum of the Arts.

Bruner, J. S. (1966). *Toward a theory of instruction* (Vol. 59). Harvard University Press.

Campanella, T. J. (2017, July 9). How low did he go? Retrieved from https://www.citylab.com/transportation/2017/07/how-low-did-he-go/533019/

Cazden, C. (1988). Classroom discourse: The language of teaching and learning (2nd ed.). Portsmouth, NH: Heinemann.

Chang, J. (2005). *Can't stop won't stop: A history of the hip-hop generation*. Macmillan.
Chen, X., Chang, L., & He, Y. (2003). The peer group as a context: Mediating and moderating effects on relations between academic achievement and social functioning in Chinese children. *Child development, 74*(3), 710–727.
Coleman, J. S. (1988). Social capital in the creation of human capital. *American Journal of Sociology*, S95–S120.
Cook, L., & Friend, M. (1995). Co-teaching: Guidelines for creating effective practices. *Focus on Exceptional Children, 23*, 1–16.
Dunley, T. (2000, May 12). The colour barrier is no more. So whose music is it anyway? Montreal Gazette, p. A1.
Easley, D., & Kleinberg, J. (2010). *Networks, crowds, and markets: Reasoning about a highly connected world*. Cambridge University Press.
Emdin, C. (2016). Hip-hop based interventions as pedagogy/therapy in STEM: A model from urban science education. *Journal for Multicultural Education, 10*(3), 307–321.
Emdin, C. (2016). *For White Folks Who Teach in the Hood … and the Rest of Y'all Too: Reality Pedagogy and Urban Education*. Beacon Press.
Emdin, C. (2011). Moving beyond the boat without a paddle: Reality pedagogy, Black youth, and urban science education. *The Journal of Negro Education, 80*(3), 284–295.
Emdin, C. (2010). *Science education for the hip-hop generation*. Rotterdam: Sense Publishers.
Emdin, C., Adjapong, E., & Levy, I. (2016). Hip-hop based interventions as pedagogy/therapy in STEM: A model from urban science education. *Journal for Multicultural Education, 10*(3), 307–321.
Foster, M. (2002). *Using call-and-response to facilitate language mastery and literacy acquisition among African American students*. (ERIC Document Reproduction Service No. ED468194)
Foster, M. (1989). It's cookin' now: A performance analysis of the speech events of a Black teacher in an urban community college. *Language in Society, 18*, 1–29.
Freedman, S. W., & Appleman, D. (2009). "In It for the Long Haul"—How Teacher Education Can Contribute to Teacher Retention in High-Poverty, Urban Schools. *Journal of Teacher Education, 60*(3), 323–337.
Fresh, M. (1984). *Breakdancing*. New York: Avon Books.
Gardner, H. (1993). *Multiple intelligences* (Vol. 5, No. 7). New York: Basic Books.
Hooton, C. (2015, July 14). Hip-hop is the most listened to genre in the world, according to Spotify analysis of 20 billion tracks. Retrieved from http://www.independent.co.uk/arts-entertainment/music/news/hip-hop-is-the-most-listened-to-genre-in-the-world-according-to-spotify-analysis-of-20-billion-10388091.htmlhttp://www.independent.co.uk/arts-entertainment/music/news/hip-hop-is-the-most-listened-to-genre-in-the-world-according-to-spotify-analysis-of-20-billion-10388091.html
John-Steiner, V., & Mahn, H. (1996). Sociocultural approaches to learning and development: a Vygotskyian framework. *Educational Psychologist, 31*(3/4), 191–206.
Kahle, J., Meece, J., & Scantlebury, K. (2000). Urban African-American middle school science students: Does standards-based teaching make a difference? *Journal of Research in Science Teaching, 27*(9), 1019–1041.

Kugelberg, J. (Ed.). (2007) *Born in the Bronx: A visual record of the early day of Hip Hop*. New York: Rizzoli International Publications.

Ladson-Billings, G. (2006). From the achievement gap to the education debt: Understanding achievement in US schools. *Educational Researcher, 35*(7), 3–12.

Ladson-Billings, G. (1994). *The Dreamkeepers: Successful teaching for African-American students*. John Wiley & Sons.

Lamar, K. (2015). Mortal Man. [Kendrick Lamar & Tupac Shakur] On *To Pimp a Butterfly* [CD]. Los Angeles, CA: Top Dawg Entertainment

Lane, C. (2008). The distance learning technology resource guide. *Retrieved, 5*(05), 2011.

Langdon, D., McKittrick, G., Beede, D., Khan, B., & Doms, M. (2011, July). Stem: Good jobs now and for the future. Retrieved from http://www.esa.doc.gov/sites/default/files/reports/documents/stemfinalyjuly14_1.pdfhttp://www.esa.doc.gov/sites/default/files/reports/documents/stemfinalyjuly14_1.pdf

Lave, J., & Wenger, E. (1991). *Situated learning. Legitimate peripheral participation*. Cambridge, England: Cambridge University Press.

Lippman, L., Burns, S., & McArthur, E. (1996). *Urban schools: The challenge of location and poverty*. Diane Publishing.

Munce, R., & Fraser, E. (2012). Where are the stem students? Retrieved from http://www.stemconnector.org/sites/default/files/store/STEM-Students-STEM-Jobs-Executive-Summary.pdfhttp://www.stemconnector.org/sites/default/files/store/STEM-Students-STEM-Jobs-Executive-Summary.pdf

Murphy, R. J., Gray, S. A., Straja, S. R., & Bogert, M. C. (2004). Student learning preferences and teaching implications. *Journal of Dental Education, 68*(8), 859–866.

National Governors' Association. (2005). *Closing the achievement gap*. Retrieved March 4, 2015. http://www.subnet.nga.org/educlear/achievement/

National Science Foundation, National Center for Science and Engineering Statistics. (2015). Women, Minorities, and Persons with Disabilities in Science and Engineering: 2015. Special Report Retrieved from http://www.nsf.gov/statistics/wmpd/

Nelson, M., & Chandler, W. (1999). Some tools common to art and science. *Art Education, 52*(3), 41–47.

Norman, O., Ault, C. R., Bentz, B., & Meskimen, L. (2001). The Black–White "achievement gap" as a perennial challenge of urban science education: A sociocultural and historical overview with implications for research and practice*. *Journal of Research in Science Teaching, 38*(10), 1101–1114.

Office of the Press Secretary (2015, March 23). *Fact Sheet: President Obama Announces Over $240 Million in New STEM Commitments at the 2015 White House Science Fair*. Retrieved from https://obamawhitehouse.archives.gov/the-press-office/2015/03/23/fact-sheet-president-obama-announces-over-240-million-new-stem-commitmen

Piaget, J. (1951). Principal factors determining the intellectual evolution from childhood to adult life. In *Organization and Pathology of Thought*, ed. D. Rapaport. New York: Columbia University Press.

Piestrup, A. M. (1973). Black dialect interference and accommodation of reading instruction in the first grade. Berkeley: University of California, Language Behavior Research Lab.

Robelen, E. W. (2011). STEAM: Experts make case for adding arts to STEM. *Education Week*, *31*(13), 8.

Schinske, J., Cardenas, M., & Kaliangara, J. (2015). Uncovering scientist stereotypes and their relationships with student race and student success in a diverse, community college setting. *CBE-Life Sciences Education, 14*(3), 35.

Smitherman, G. (1977). "Talkin and testifyin: The language of Black America." Detroit: Wayne State University Press.

Tate, G., & Goodison, C. (2012). An interview with Greg Tate. *Callaloo, 35*(3), 621–637.

U.S. Department of Education, National Center for Educational Statistics. (1996). Urban schools: The challenge and location of poverty. Washington, DC.

CHAPTER THREE

Seeding the Future

Social-Justice Driven STEM Education

CHRISTIAN KONADU ASANTE,[1] JACQUELINE DELISI,[2]
MEGAN McKINLEY,[3] AND MICHAEL BARNETT[4]

Abstract

There is significant, widespread commitment to increasing the active engagement and representation of non-dominant student populations, namely from low-income communities and communities of color, in STEM. In this chapter, the authors (1) describe a framework for social-justice-driven STEM (STEMJ) curriculum used in science classrooms in school and in an out-of-school time (OST)

---

1 Christian Konadu Asante, Doctoral Student
   Lynch School of Education, Boston College
   asantec@bc.edu
2 Jacqueline DeLisi, Research Scientist
   Education Development Center
   jdelisi@edc.org
3 Megan McKinley, Doctoral Student
   Lynch School of Education, Boston College
   mckinlmb@bc.edu
4 Michael Barnett, Professor
   Lynch School of Education, Boston College
   barnetge@bc.edu

program for underrepresented students involved in a range of interdisciplinary learning experiences including urban farming, solar energy, coding and robotics; (2) explore how participation in the OST STEMJ program impacts youth's beliefs about STEM, the role that STEM has in their lives, and their understanding of the role of failure in the scientific process; and (3) explore the experiences of teachers as they implement the social-justice-driven STEM curriculum in their classrooms.

Editorial Reflections

We appreciate the upcoming Seeding the Future chapter since it involves both students and teachers. The chapter feels like a work in progress and so has research questions interspersed throughout its text. It also weaves together in-school and out-of-school learning, which we define as a STEM Ecosystem. These connections are especially important for lower income students since they are the most likely to A) receive lower quality education, AND B), have limited access to high quality out of school learning experiences. If you have ever been to the Northeastern part of the United States, it won't be hard to imagine rowhome landscaping—small "green" spaces bounded by slabs of concrete only a few feet away from the street, an alleyway or a set of stairs. The space in front of my, Joy's, humble graduate school seemed a perfect lab space for me to build a community garden with my son and the children who lived next door. In a back "yard" plot, I had composted for years throwing fruit and vegetable food waste into the little box just outside my kitchen. For years, "opportunity" harvests produced tomatoes that we enjoyed without a whole lot of effort. The hearty seeds grew freely. The more delicate herbs and vegetables I planted with intention always seemed to fall short of the tomatoes' glory. The neighbor's children became interested in my garden when their ball flew over the fence. They were fascinated! To help build interest and perhaps stimulate a sense of neighborhood pride, I invited them to help me compost in one of the two small soil squares we shared in the front of the house. We did. I could see the soil become darker and more enriched. The children lost interest until what seemed like overnight a vine grew but only one fruit was born. I had no idea what it was so I watched in equal anticipation. After a few weeks, I realized that it was not a cucumber but a watermelon! The whole neighborhood watched as it grew. Once it got to be about 10-inches long, I harvested it ... prematurely. It was SO funny! This fruit had grown from the seeds of a watermelon that I had dropped trying to carry it into the house.

Composting is such an easy action to take to address urban problems. Food in/security aside, having a way to restore nutrients to soil battered by erosion and overuse is an easy connection to make. The exercise of treating composting as a community experiment involving more than one household and generation is equally promising for urban environmental justice. Looking back on it now, I think about the lessons I learned by composting in front of and behind my house, and smile. For now, I will leave agriculture to the experts and support local farms. Filling opportunity gaps with programs that promote the social justice in STEM (STEMJ) notion of "out of school time" activities tied to food security and urban farming is an important example of transgressive teaching. As communities work to reconcile a lack of choice in the marketplace, it makes perfect sense to build agency with programs that start by teaching how to meet basic human needs.

My, Janelle's, story involves another STEM ecosystem that has shown promise for transformative outcomes for its student participants. I work with an organization called GLOBE—Global Learning and Observations to Benefit the Environment. As a GLOBE Partner, I had the opportunity to fund one group of student scientists and their teachers to attend the first ever regional Student Research Symposium (SRS), to be held that year in Houston. I put out the call for teams across the Southwestern US, but the timeline until the SRS was tight. The teams would have to do the data collection, analyze the data, create research posters, and get permission from their schools and their families to travel. I was thrilled when one team responded, and cognizant of opportunity gaps, even more thrilled when I learned that all the students and one of the two teachers were Latinx.

At the student research symposium, each team presented their work to panelists who were working scientists. Many of the other students at the symposium had far more experience with this type of work, including knowing how to create professional caliber posters, but the Colorado group did well, and they viewed the whole event as a learning experience. I felt lucky the group was even able to attend on such short notice, and I was certainly impressed with all the work the teachers had clearly done over time. When I asked the two teachers how they had been able to do this work so successfully with the kids, they described to me the different expertise each of them brought to the work. The White teacher previously worked as a scientist, and had terrific content knowledge and passion for STEM. The Latinx teacher, who grew up in the school community, told me she actually used to hate science and had been tracked out of it in her own education. She talked about the close relationship she had with the students and their families. The White teacher explained that the Latinx teacher knew how to communicate effectively with the families, and that she understood and addressed any of the families' doubts about both in-school and out-of-school STEM learning. Though the teachers would not necessarily describe their work as being motivated by social justice, the work they are doing supports transformative outcomes for youth of color in STEM. Similar to the work of the Seeding the Future project, the teachers are working in a STEM Ecosystem that blends in-school and out-of-school learning in a way that has positively shaped students' peer to peer discourse. These efforts to build teacher and student capacity to open STEM career pathways represent promising cases of the Opportunity to Learn.

Introduction

Our work focused on the recruitment of low-income youth from populations that are underrepresented in science into our program where social justice concerns (food justice, food security) are illuminated, analyzed, and acted upon with and through the development of STEM knowledge and skills. Unlike many programs which focus on teaching STEM to close the opportunity gap, our program recognized the potential for urban youth to become deeply knowledgeable citizens who understand the localization of food injustice within their communities and can mobilize their enhanced STEM knowledge and skills to illuminate and resolve social injustices,[1] such as food deserts. In this chapter, we will (1) explore how participation in a social justice-based out of school STEM program[2] impacts youth's interest in STEM and intentions of pursuing a career in STEM and problem solving (2) explore the experiences of teachers as they implement the social-justice driven STEM curriculum in their classrooms.

Contexts of the Work

During the in-school program, students use hydroponics systems to grow food such as basil, lettuce, kale, or tomatoes in the classroom. Hydroponic systems allow students to be immersed in interdisciplinary STEM as our hydroponic program requires that students learn and apply aspects of chemistry, physics, biology, economics, and inquiry. To implement hydroponics in their classrooms, we provide teachers with two kinds of hydroponic kits. The first is a nutrient film technique, and the second is a "float" or Deep Water Culture system. In the nutrient film technique system the water is continuously recycled and flows as a thin film on the bottom of the growth channels. The water is continuously recycled with a water pump, which supplies the water with sufficient dissolved oxygen levels for the roots to uptake the nutrients. In the Deep Water Culture system, the plants float on top of the water with their roots in water that is oxygenated with an air pump.

Our program allows the teachers the flexibility to engage their students in putting together the hydroponic system. Once the systems are constructed and the initial crop of seedlings are planted, there is little maintenance required unless a problem arises with pH, lighting, or nutrient concentration. Therefore, while the first crop grows, our curriculum focuses on understanding the systems and conducting scientific research by asking questions regarding the impact of different variables on food production. In this way, teachers are able to engage students in inquiry and the design of experiments, collecting data while the plants grow, and analyzing results in depth when the plants are harvested. The daily or weekly data collection encourages careful observation of the changes that take place as the plants grow. Students in our program have been consistently excited to see the changes happening in what they come to see as "their" plants.

We have learned that hydroponics is flexible for integration into most teachers' curriculum because it utilizes principles and concepts from a range of disciplines. For example, some of the variables that impact plant growth in hydroponics systems are related to physics, including types of light (LED, high intensity, sunlight), the electrical conductivity of the nutrient solution, and the amount and rate of flow of; some variables are related to chemistry, such as the pH of the nutrient solution and the composition and concentration of nutrients; and some variables are connected to biology, like the light or nutrient needs of different types of plants and the impacts changes in these variables have on plant health.

Finally, when the plants are fully grown, students harvest the crop and begin a new batch of seedlings. The growing times of foods are typically shorter in hydroponic systems than in soil-based agriculture–lettuce, for example, can be harvested after three weeks–allowing for multiple harvests, and multiple rounds of experiments, throughout the school year. Students in our program have eaten the plants in class, shared them with their school cafeteria, sold them to the school or local community, or taken them home to eat with their families. For some, the need to deal with the produce at the end of each harvest has led to connections beyond the classroom. In the following we present two classroom cases where teachers are using hydroponics to engage their students in learning science as well as reaching out to the larger community.

The out-of-school time (OST) program also engages student in activities such as building circuits, connecting solar panels, and conducting air quality assessments. The curricular framework used in our OST STEM program draws upon Bronfenbrenner's[3] notions of meaningful learning in which individuals are guided to develop a comprehensive understanding of the various aspects of their environment that they affect and are affected by.[4] College Bound is a collaborative venture between teachers, students and peers as they work together on a social-justice driven STEM project. The program takes place within a college readiness program for students in grades 7–12. The general goals of the program include increasing interest and persistence in STEM, supporting critical mindfulness and problem solving, and increasing self-efficacy in applying to and persisting in college. Students participate in the program throughout the school year in addition to a two-week summer camp. During the school year, students meet every other Saturday from 8:30 AM–3:00 PM and participate in three sessions: social justice driven STEM learning (STEMJ), college and career planning, and social justice action projects. Within the STEMJ sessions, students are members of one of three groups including air quality monitoring, building electric circuits and hydroponics. All groups play a role in working toward a common goal of growing vegetables and fruits to sell at local farmers' markets.

Research Methods

The Seeding the Future in school project introduces classroom teachers to hydroponics content and materials through a Summer Institute and provides school-year support for the implementation of the materials in their classrooms. This project is part of a larger body of work that also includes bringing hydroponics and STEM social justice curricula and materials to educators and students in the out-of-school College Bound program and in other informal learning environments. The College Bound program takes place on the campus of Boston College. It is a STEM-social justice curriculum that recruits students from the Boston Public School system.

In-School Study

The questions addressed regarding the in school work with classroom teachers include (1) What are the challenges and successes that classroom teachers anticipate in using the hydroponics materials in classroom? (2) In what ways does the Summer Institute help to prepare teachers for implementing the materials in their classroom? (3) What is the potential of the curricular materials and teacher supports for influencing teachers' use of inquiry-based instructional methods, their comfort using hydroponics materials, and their science self-efficacy?

In order to address these questions, we conducted focus groups with the classroom teachers during three years of the Summer Institute. We also administered a survey to teachers about their school contexts, science self-efficacy, and instructional practices. STEM Self Efficacy was determined using the *Science Teaching Efficacy Beliefs Instrument*[5] and consists of two scales: Science Teaching Efficacy Beliefs (STEBI, 11 items) and *Science Teaching Outcome Expectancy* (STOE, 9 items). Teachers' reported comfort with teaching hydroponics consisted of one item that asked them to rate their comfort on a five-point scale, five meaning extremely comfortable. Items that indicated their use of open-ended and inquiry-based methods were adapted from the *Teacher Efficacy and Attitudes toward STEM Survey* (T-STEM).[6]

Teachers who attended focus groups each year varied in the grade level taught and their roles in the school, with most working in middle or high schools and many teaching in schools or classrooms with high proportions of students with special needs or English Language Learners. Teachers also represented a variety of roles in schools, including a school administrator, a health teacher, and a math specialist. In each year of the focus groups, about half of teachers described previous experiences using hydroponics or gardening activities with their students, either through work with the BC program or through their own initiative. One advantage of the hydroponics program is the number of connections that can be made to students' learning across grade levels and varying abilities, and to their varied school settings. In addition, many teachers described their implementation of hydroponics in terms of the needs and intended outcomes among their population of students, including English Language Learners (ELLs) and students with special needs.

Teachers were also prompted to describe challenges for implementation, and some cited the need for space or challenges in talking to administrators or custodians. In some years teachers described the needs for sufficient electricity to maintain the hydroponics systems. However, teachers who had multiple years of implementing the materials stated that their administrators and schools are supportive in part because the materials are aligned with the standards they need to address. Our survey data indicate that teachers who had attended at least some hydroponics professional development reported slightly higher science self-efficacy and higher rates of inquiry-based instructional practices.

OST Study

A total of eight students between the grades 8–12 were involved in the OST portion of the study. The research was conducted on the campus of Boston College as part of a broader out of school program called College Bound. Participants in the program included students in grades 7–12, who are predominantly Black, Latinx, and Caribbean migrants. In terms of socio-economic status about 78% of students qualify for reduced-price or free lunch. Our recruitment targeted students with average academic performance but with identified potential to excel if given the opportunity.

We used semi structured interviews which allowed us to adapt our research questions to the interview sessions. The research questions investigated during the interviews are (1) How does College Bound facilitate students' interest in STEM? (2) To what extent does the College Bound intervention foster progress in students' intentions to further pursue STEM educational options and to consider STEM careers? (3) What are students' experiences with problem solving in College Bound? The interviews were conducted in the summer of 2012. The responses were audio recorded and later transcribed.

Project Outcomes

After thorough readings and discussions among researchers, themes related to our research questions were identified, which formed the basis of our analysis. Data were collected from teachers and students. Outcomes for the research are reported here.

Teacher Outcomes

Several implications emerged from our data on classroom teachers' use of hydroponics activities. First, following the Summer Institute the teachers were satisfied with the materials and curricular supports provided by the project. Attention to the connections to standards in the materials as well as suggestions for applying the activities to a variety of content areas made the materials particularly appealing. Second, across the group of teachers, the hydroponics activities have been and will be used with a variety of students and school settings, from younger elementary aged students through high schools, across a variety of STEM and other disciplines and with students, such as those who are ELLs or who may not traditionally have access to hands-on and engaging STEM activities. Finally, teachers who returned to the Institute for a second year or who had previously implemented hydroponics activities with their students were particularly mindful of the ways in which they could use the activities with their students. The returning teachers provided insights into the kinds of supports that might be needed. For example, as teachers work to overcome the challenge of implementing hands-on and engaging inquiry-based activities it may be important for them to be mindful of developing cross-curricular connections and partnerships with other teachers and departments in their schools.

The goal of our research into the teachers was to inform the materials and professional development by understanding teachers' perceptions of the successes and challenges they face. Since our data were collected following the teacher institute each year, it will be important to follow up with teachers in the future to understand their implementation of the hydroponics activities in more detail. Further study could yield insights into the classroom implementation and more specifically the ways in which the hydroponics materials can be used to develop and support students' understandings of science and engineering practices.

Student Outcomes

Interview responses from the students revealed that six out of eight students involved in the study had indicated an interest in STEM after the program. One example is Student G. When asked how engaging in the program has influenced his thinking about STEM, he reported:

> Well, I didn't used to like science, but my mind changed over time when I felt like it was getting hard. But then in math, I think everything you had to do, like, basically involved math. So, it's all, so, when I came here, like, doing all this stuff about pints and other stuff, I became to realize that, like, I don't know, I completely changed my mind about science even though it might be hard. But I think the apps [inaudible] and getting involved in what the teacher might be doing might get me, you know, to like it even more.

It is apparent that Student G has developed an emerging interest in STEM based on engaging in our summer hydroponics project. He makes brief references to how the program structure and various activities got him interested in STEM. He continues:

> Building the things to MST, yeah, I think, when you build and create plants without using food. And I think it's really interesting and getting involved with other people and then explaining how, like, you get through work, how like you can, like, make plans without like having big garden. You can, like, do this at your house. It's good. Well, it's like at first, when they started talking about how like you couldn't use, like, these, like, you know, how when you usually plant like seeds, for them to find another way without using seeds, I thought was interesting.

There is evidence of an emerging interest in STEM based on his statements, but also the importance of interacting with his peers, negotiating, and applying how the hydroponics project is useful outside the setting in which the program took place. Student G stated that "You can like do this at your house. It's good." He seems to be making connections between the hydroponics project and what can be done outside. The peer interaction and working together as team might have influenced his interest in STEM, which previously existed to a lesser degree.

Based on such responses, we observe that the social interaction among peers, problem solving, and connections to events at home are all extended to STEM. These experiences have presented a new way of thinking about STEM. Another feature of the program that got students interested in science was leadership, mentoring, and the hands on activities. For instance, Student D made connections to STEM based on the mentoring activities he participated in, or "my relationship with them and watching them." He also expressed a new interest in STEM based on how extensive and practical the program was. Similar to Student G, Student D also made connections to STEM as a result of participation in the program:

> Um I think it changed in science because when I was in 8th grade, I really didn't like the science subjects that they were doing. But then when they taught me about the air pollution and solar panels it was really engaging.

This student makes a subtle distinction between science at school and science he engaged in during the program. He prefers the science during the program for its practicality and possibly because it is quite different from science done at school. We are encouraged by student views and responses because the trends and patterns they have observed, such as team work with their peers, hands-on activities, mentorship, usefulness to the home, and leadership, are all important ingredients for many professional careers.

Students' STEM Career Interests

Students expressed their STEM career interests around specific activities such as building circuits, and they made connections to job opportunities. Student E indicated an interest in a career in STEM as a result of certain activities in the program such as building circuits:

> My favorite part was making the circuits because it was something I learned in 3rd grade. And I never, like, in all my years of school after, I never really learned more about it. So, going back to it really made me, like, remember, and it's something I could use in my future.

This student recognizes that building circuits has connections to a future career in STEM, specifically software engineering: "Yeah, well because I wanna be a software designer and I can use circuits because software is made of electricity. I think I could make more complex designs." His decision was also influenced by another important part of the project, which is the social justice component, where students were treated to a series of seminars on jobs and finances and earning opportunities. It is noteworthy that Student E is able to establish the relationship between STEM and job opportunities and money:

> Like, in a class we would see, and they would say, like, they would show us, like, a salary, like, what we would need and stuff. So, it made me think more realistically. So, it made me think of, like, software design is actually a little better. Like, it has a better salary and stuff.

Another participant, Student I, established a connection between the circuit building and the desire to be an engineer and attend the Massachusetts Institute of Technology (MIT). This student can be said to have successfully negotiated a

STEM Identity, with interests that have also been influenced by his parents. Student I responded:

> When we were, like, setting up circuits, there was a time before we started doing stuff with solar panels: the big ones we had ordered, there was. So, we would have a battery and the resistor or the light bulb that we had. We would connect the light bulbs to the battery, circuits. I think I've learned more about like engineering and, like, technology based stuff. It's, like, changed what I think I would want to be in life, cause, like, my mom and my dad, they want me to be, like, an engineer goes to MIT. Like, I visited MIT, like, last summer.

For students such as Student I, programs like College Bound already reinforce their conceptions of STEM and help build their confidence around long-term career interests in STEM.

> Well, I already knew before I came, I already knew that math and science were, like, my best subjects. Cause I already knew it, too, a lot of math and science already, like, for engineering. So, that's, like, helped me. I mean, get more of, like, this program has helped me get more of, like, a grasp of those subjects.

These responses and observations from students give credence to our research findings about how activities such as building circuits, solar panels, and studying hydroponics provided a space that satisfies students' learning preferences, and also provides an opportunity for students to think about and consider possible careers in STEM.

Students' Experiences with Problem Solving in College Bound

The out-of-school time program provided students with opportunities to use science to analyze real world problems and develop solutions through a social-justice lens. For example, as mentioned previously, students investigated the locations of food deserts, or the lack of access to fresh produce, which are prevalent in low-income areas. To combat this community-based issue, students developed hydroponic systems fueled by alternative energy (e.g., solar, wind) to grow produce year-round. Through this process, students not only developed STEM skills but also had the opportunity to encounter and overcome challenges. Importantly, students learned that making mistakes is inherent to successful endeavors, such as creating a functional solar-powered hydroponic system. For example, one group of students made a mistake when building their hydroponic unit, which spurred them to rebuild the whole system. A group member described this experience:

> I got more than halfway through it and then we made, like, a slight mistake … and we had to take down everything and rebuild it again. … Yeah, cause I pictured it in my

head, and when I look at it, it doesn't look like how I seen it in my head. I have to do it over until I get it right.

The student went on to reflect on the value of making mistakes:

> [Making mistakes] is valuable because then you learn and how to fix it. If it wasn't broken then you wouldn't know what to do with it. And that's how I've learned the most usually [by making mistakes]. When I do something the first time it's never right, and I have to redo it again. So, I kinda like it because I focus more. And while I'm redoing it I get, like, other ideas, and I change it up. So, the end result is even better than what I started off with.

In this statement, the student explained that making mistakes is not only valuable, but also vital to improving the final product and to learning in general. Through their participation in Seeding the Future, students directly experience the process of problem solving in STEM, which, in turn, may normalize the possibility of making mistakes or "failure" and inspire them to take more risks in the future. Students reported overcoming challenges through trial-and-error, teamwork, and support from their instructors. One student attributed their team's success to teamwork, stating, "We're doing teamwork, and we're working together. [And it] is working out cause when one person doesn't know what to do the other person kinda helps each other out." Another student expressed that same desire and described how continuously "trying" are the keys to success:

> I just realized, though, they have to want it enough in order for them to actually wanna succeed. ... It's more or less about trying. And the more you try, the easier it will get to you, and that's what I feel people should do. Again, the emphasis on continuously trying suggests that this student has realized that one's success in STEM does not happen easily and instead requires overcoming challenges and setbacks.

When students were not directly involved in designing and creating devices, such as solar-powered hydroponic systems, they engaged in problem solving by researching, analyzing, and discussing environmental issues. For example, students investigated air pollution data from various locations around the world and researched factors contributing to air pollution in these areas (e.g., population, energy use, sources of fuel). Students collaborated with their group members to analyze the data and develop potential solutions to reduce air pollution in these areas. The STEMJ instructors played an important role by probing students with questions to make them "think more deeply" about the problem throughout process. Students in the program also learned about the complexities of environmental problems that do not always have simple solutions. For example, one student expressed a desire to reduce the air pollution in China; however, after researching and discussing the problem in the program, they realized that reducing air pollution would increase the price of products imported from China. They discussed this dilemma below:

> One problem that I learned that was the biggest for me, that I really liked, was "Why can't we help China with their air pollution?" But, if we did that, then prices would be higher. Prices are already really high, and most things are made out of China. I wanted to keep prices as-is. … Some people would buy it [the products], but some people would not. Cause, I mean, some people can't buy it now cause, um, prices are good, I guess.

In this statement, the student recognized that reducing air pollution in China is a complex problem that could impact a variety of stakeholders, including consumers in the United States, in different ways.

Lastly, the College Bound program provides students with a unique opportunity to develop STEM expertise through exposure to subjects that they may not encounter outside of the program, such as alternative energy, hydroponic farming, and air quality. For example, a student describes their first experience in building a solar panel within the program stating, "One of the problems was trying to learn how to actually use a solar panel because it was my first time actually using them. It was hard at first, and I did not know what to do." Furthermore, students develop confidence in their STEM abilities through participation in the program. For example, one student describes their feeling of inadequacy in unfamiliar STEM fields stating, "If you don't know anything about it, you [begin to] question yourself, like, 'How do they do that?'… Yeah, question if you would be able to do it." They have learned to overcome this challenge within the program by "asking questions and getting involved [and] talking to people around you [to figure the problem out]." By exposing students to a variety of subjects and providing them with opportunities to become STEM experts within a social justice-driven project, students may not only consider potential STEM careers they might not have encountered otherwise, but also better understand the relevance of STEM to solving problems in their communities.

Discussion

In the United States, the representation of non-dominant students in STEM, specifically Latinx and Black students, is low compared to White and Asian students.[7] Our work attempted to fill in the gap by designing an out of school STEM program in tandem with social justice issues that targeted the above populations. The students involved in our program and subsequently our research were from non-dominant populations such as Latinx, Caribbean, and African American populations. It is notable, however, that based on student responses, our program has facilitated an interest in science and mathematics. Students demonstrate this facilitation by their responses noted above with respect to social interactions with their colleagues, usefulness of the project, and job opportunities. A similar observation by Mark et al. (2013) was made from a comparable demographic about how informal environments help facilitate interest in STEM.

Science education is a civil right.[8] Our results based on some student responses after the program indicates a renewed interest in a career in STEM. Similar results were reported[9] during an out of school STEM program: exposure of middle school students and high school students to STEM activities outside school hours is a motivating factor for subsequent enrollment in STEM programs at the college level. Our work is more detailed in that we observed and investigated the specific STEM activities and the characteristics that got students excited about science whether hydroponics, air quality or building solar panels and circuits. The responsibility rests on policy makers, researchers, practitioners to provide opportunities that continue to provide such valuable experiences for students to establish the relationship between these activities and career goals.

Finally, by observing engagement of students in problem solving activities, we hope to not only have helped students develop interest in STEM and STEM careers but also that students will have appreciation of how problems are solved in the real world. By engaging them in hands on activities that heavily involved teams, our work facilitated students' understanding of the benefits of working together as team, interacting with peers, and negotiating to get the work done. This facilitation is evident in some student responses, evidence that engaging with peers, kinship, and leadership around a specific scientific activity are vital instruments in facilitating students' interest in science and attitudes towards learning.[10]

Importance and Contribution

Over the past decade there has been an increased interest in embedding more explicit STEM career development into science education for youth,[11] yet a key question has remained unanswered. Specifically, how can we move from helping students explore STEM fields to having them actually consider STEM fields as viable career aspirations? To answer that question, we believe that engaging youth in using science to solve social justice related problems, while explicitly supporting them in analysis of how the skills they learn while solving those problems are related to their future goals, is critical to engaging youth in not just learning STEM but also sustaining interest in STEM.

Discussion Questions

- What are some effective practices you have found to create a safe space for students to make mistakes?
- What is the role of "mistakes" in the STEM equity paradigm?
- Which aspects of hydroponics and urban farming initiatives drive Social Justice STEM education (STEMJ)?
- What social justice issues relevant to your students could you utilize as a basis for an integrated STEM unit?
- In what ways does teacher self-efficacy in STEM (or a lack of it) shape student learning in your context?

Notes

1. Tan & Calabrese Barton, 2010.
2. OST STEMJ by Madden, Wong, Vera Cruz, Olle, & Barnett, 2017.
3. Bronfenbrenner, 1994; 2009.
4. See http://edibleboston.com/boston-teens-grow-green/ and http://www.wcvb.com/article/thursday-august-25-farm-finds/8246770
5. Riggs & Enochs, 1990.
6. Corn et al., 2013.
7. NSF, 2017.
8. Moses & Cobb, 2001.
9. Dabney et al., 2012.
10. Basu & Calabrese Barton, 2007.
11. Connors-Kellgren, Parker, Blustein, & Barnett, 2016.

References

Basu, S. J., & Calabrese Barton, A. (2007). Developing a sustained interest in science among urban minority youth. *Journal of Research in Science Teaching, 44*(3), 466–489.

Bronfenbenner, U. (2009). The Ecology of Human Development. Cambridge, MA: Harvard University Press.

Bronfenbenner, U. (1994). Ecological models of human development. In *International Encyclopedia of Education* (pp. 1643–1647). Oxford: Elsevier.

Connors-Kellgren, A., Parker, C. E., Blustein, D. L., & Barnett, M. (2016). Innovations and challenges in project-based STEM education: Lessons from ITEST. *Journal of Science Education and Technology 25*(6), 825–832.

Corn, J., Bryant et al. (2013). Second Annual Race to the Top Professional Development Evaluation Report: Part II Local Outcomes Baseline Study. Raleigh, NC: Friday Institute for Educational Innovation, North Carolina State University. Retrieved from http://cerenc.org

Dabney, K. P., et al. (2012). Out-of-school time science activities and their association with career interest in STEM. *International Journal of Science Education, Part B, 2*(1), 63–79.

Madden, P. E., Wong, C., Vera Cruz, A. C., Olle, C., & Barnett, M. (2017). Social Justice Driven STEM Learning (STEMJ): A Curricular Framework for Teaching STEM in a Social Justice Driven, Urban, College Access Program. *Catalyst: A Social Justice Forum, 7*(1), 4.

Mark, S., Debay, D., Zhang, L., Haley, J., Patchen, A., Wong, C., & Barnett, M. (2013). Coupling social justice and Out-of-School time learning to provide opportunities to motivate engage and interest under-represented populations in STEM fields. *Journal of Career Planning and Adult Development, 29*(2), 93–105.

Moses, R. P., & Cobb, C. E. (2001). *Radical equations: Math literacy and civil rights.* Boston, MA: Beacon Press.

National Science Foundation [NSF]. (2017). Women, Minorities, and Persons with Disabilities in Science and Engineering: 2017 (NSF 17-310). Retrieved from https://nsf.gov/statistics/2017/nsf17310/static/downloads/nsf17310-digest.pdf

Riggs, I. M., & Enochs, L. G. (1990). Toward the development of an Elementary Teacher's Science Teaching Efficacy Belief Instrument. *Science Education, 74*(6), 625–637.

Tan, E., & Calabrese Barton, A. (2010). Transforming science learning and student participation in sixth grade science: A case study of low-income, urban, racial minority classroom. *Equity & Excellence, 43*(1), 38–55.

SECTION TWO

Engagement

Extended Learning Opportunities

Section Two of the volume focuses on Engagement. Keiler and Robbins' Chapter Four analyzes a program that works with both students and teachers, and specifically examines the nature of relationships between students and teachers. The program leverages peer leadership by middle performing students to transform classroom environments and student outcomes in STEM classes and dramatically improve college readiness. Teaching Assistant Scholars serve as peer instructors and become more effective students themselves through a specially designed course, while teachers lead and assess an instructional team that enables student success. A two-year professional development model supports teachers as they shift their practice and develop new skills in order to facilitate positive STEM learning experiences for all students. Students enter a pipeline to college that consists of internships, advanced courses, and bridge to college mentoring.

In Chapter Five, Xu, Newton, Turrin, and Vincent theorize strategies that have been effective in engaging non-dominant students in a summer field research experience. The authors present a science mentoring practice deeply rooted in two distinct local cultures: the Lamont-Doherty Earth Observatory (LDEO: a globally-recognized earth/environmental science institute), and mostly poor and working class NYC public school students. The program was developed slowly over time, becoming larger, more institutionalized, and sustained with support from private foundations and federal agencies. Program outcomes are deemed successful in meeting its mission, with 100% of alums college bound and more than 40% choosing to study in STEM fields.

Bhaduri, Gendreau, Koushik, Ristvey, Russell and Sumner analyze an afterschool program that provides engineering experiences for middle school students in relatively lower socioeconomic contexts in Chapter Six. Engineering Experiences is an afterschool project to understand and promote practices that develop students' motivations and capacities to pursue careers in STEM fields by providing engineering experiences in atmospheric sciences and related areas to middle school students. This collaborative project is working with low-income youth and combines hands-on learning in STEM with engineering experiences and ongoing mentoring to prepare students for success in high school coursework and the future workforce. This article reports the ways such experiences influence underrepresented students' competency, motivation, and persistence in STEM.

CHAPTER FOUR

New Roles and Relationships in Urban STEM Learning Environments

How the Peer-Enabled Restructured Classroom Enhances Equity and Access

LESLIE S. KEILER[1] AND KATHLEEN ROBBINS[2]

This research was supported by NSF grant #DUE-1102729

Abstract

The Peer Enabled Restructured Classroom (PERC) model radically alters interactions among students and between teachers and students in urban STEM classes. The PERC model places average-performing students as peer instructors called Teaching Assistant Scholars (TAS) leading collaborative learning teams, leveraging the power of peer-to-peer motivation, and scaffolding

---

1   Leslie S. Keiler, Associate Professor of Teacher Education
    York College, The City University of New York
    lkeiler@york.cuny.edu
2   Kathleen Robbins, Current Teacher
    Bronx Early College Academy of Teaching and Learning
    loveflyfishing@aol.com

to create student-centered learning environments. The main objectives of the PERC model are to close gaps in student outcomes of STEM performance, high school graduation rates, and college readiness. High stakes test results show consistently higher outcomes in PERC classes compared to traditional classes in urban high needs schools. In PERC, the teacher begins class by providing a conceptual context, then at least two thirds of each class each day is spent in small group work led by TAS. In a well-implemented PERC class, students stay on task and discuss STEM concepts, guided by TAS. They follow instructions provided by TAS, ask TAS questions when they are confused about content or task requirements, answer scaffolding questions posed by TAS, and have TAS check the quality of their work before moving on to a new task. Teachers act as managers of their instructional team and assessors of understanding. These altered roles result in new relationships among students and between students and teachers, empowering students and altering teachers' expectations. In this chapter, these new classroom roles and relationships are explored and analyzed to explain the model's outcomes.

Editorial Reflections

One of the jobs teachers have is evaluating success. Is the alternative to success always failure? Like many school communities are working to address the participation gap that persists for students in STEM. I, Joy, work in a system grappling with ways to support students missing from advanced STEM classes. There are a few who are willing to step outside of their comfort spaces and try advanced classes, but retention rates are low for Black and Latinx students. Even when students enroll in spring for courses that begin in September, they often develop exit plans before winter. To address this issue, my district developed a summer preview course to expose students to content and "success" strategies. I helped identify students who should take the summer course. I actually did not recommend that they take the full-year course; I just wanted them to develop social strategies of collaboration around academic activities with other students who could effectively lead them. These were the students who were leaders in my classes but who had other factors working against their full participation in STEM. Many were multilingual, eligible for free and reduced lunch, and were used to being in low-tracked classes. Grounded in the Aspirations framework, my overarching goal in recommending them for the class was to help them develop agency and resilience. The problem, however, was measuring success. We did not have an adequate resource for evaluation. I developed the "Chemistry Learning for Academic Success in Science" (or Core/Chemistry Lab and Academic Skills for Success in Science) rubric (see Appendix

II) as a means for addressing this missing link in the conversation. The purpose of the C.L.A.S.S. rubric was to identify core skill acquisition with two goals in mind: to provide a basis for student's personal reflection and to identify qualitative measures of potential success. As a tool for low-stakes evaluation, I think it works. This is a work in progress for my professional learning community. Even this is part of my story: to see almost everything as an action research opportunity, examining what we need to do differently. Equity-minded teachers constantly interrogate their practices and the tools they use to communicate with students, taking a step to the side to learn like students and make space for students to teach.

I, Janelle, work with both preservice and in-service teachers. Most of them look like me—White and middle class, though many of the students at my urban commuter university have far less socioeconomic privilege than I have had throughout my life. I always struggle with how to prioritize preparing them as best as I can for the classrooms they will encounter. I want them to understand the macro nature of the schooling system in which they will work, and I also want them to have a sense of agency in the micro levels where they move. I try to help them discover their own biases, and to realize that that process will often be very uncomfortable, especially if it's not something they have done before. They will not be able to build authentic relationships with students from different backgrounds than theirs if they don't undertake this work and realize that it must be ongoing. Equity metrics are not to be seen as an add-on, but rather as a central aspect of their teaching practice, interrogating what they must do differently in order to better serve their students. We know that students must be engaged and motivated if learning is to occur, and that peers are an extremely strong influence on student engagement and motivation. That influence can either support or hinder school-based learning. How do teachers who are not members of their students' communities tap into that influence as a resource to facilitate learning?

The demographics of the preservice teachers at my university reflect larger trends of heavily skewing White, despite our institution's relative racial, ethnic, and linguistic diversity. These numbers tell the field of teacher education that we have a lot of work to do. There are many of us who are trying different avenues—actively working to recruit more students of color into teaching; taking on additional advising both formally and informally; working to hire more faculty of color to the department so our professors more closely mirror the community; and working to build supportive structures based directly on what students report as institutional roadblocks.[1] The upcoming chapter on Peer Enabled Restructured Classroom describes a set of very promising structures that provide previously marginalized students with Opportunity to Learn. The team's work to cultivate students' leadership skills and peer interactions within STEM learning contexts represents transgressive teaching, inviting other teachers and teacher educators to rethink our own structures.

Introduction

The Peer Enabled Restructured Classroom (PERC) transforms urban science classrooms into student-centered, peer-facilitated learning environments where students actively engage in academic work and develop skills essential for college success. Funded through several grants from the National Science Foundation, a partnership of high schools, higher education institutions, district administration, and school support organizations developed, piloted, and revised this model to revolutionize STEM education in high needs urban schools. The PERC model restructures STEM classrooms using the resources readily available in schools—students and teachers—making the model low-cost and sustainable. This chapter focuses on the experiences of teachers and students in the PERC Living Environment (LE) course, a biology-based course that ends in a high stakes state Regents exam generally required for high school graduation. In PERC classes, a team of average-performing students, under the management of a trained PERC teacher, lead learning teams through academic tasks for the majority of each class period every day. The peer leaders are students who succeeded at a marginal level in the course the previous year, passing the high stakes exam but not at a level regarded as college ready. They are called Teaching Assistant Scholars (TAS) in recognition of their dual role of supporting peers in PERC classes and progressing in their own academic development and college-readiness. The TAS take a separate class each day, taught by the PERC teacher, in which they learn how to succeed in these roles. The initial NSF grant funded the development of the PERC instructional model, focusing on the experiences and outcomes of the students taking STEM courses for the first time. A subsequent grant was used to explore and expand the potential of the TAS, developing the TAS course and a possible pipeline into college. This chapter documents the experiences of the students, TAS, and teachers during the initial years of this focus on TAS, recognizing this as a unique contribution of the PERC model.

The partnership developed the PERC model with the essential objectives of improving STEM performance among historically under-performing groups, raising graduation rates at high needs schools, and increasing the number of graduates ready to succeed in college. Mirroring what was then a national focus on college readiness,[2] New York City schools have been evaluated and rated based upon their ability to prepare students for college and career success, as measured by graduation rates for different levels of diplomas and proficiency in reading, writing and mathematics, including placing out of remedial college coursework.[3] College admission and success depends upon academic competency in high school science, which remains a national concern.[4] However, K–12 students' scores on tests of conceptual understanding and application of science remained low, remaining stagnant at the high school level, with few students performing above proficient on national measures.[5] Further, performance gaps in science persist between White students and Black and Latinx students, as well as between those students who qualify for free and reduced price lunch and their more affluent peers, although there is some evidence for slight narrowing of the gap for non-dominant school students.[6] New York State science scores have continued to stay flat and slightly below national averages, while national figures have shown minor improvements.[7]

The PERC model emerged out of collaborative efforts to improve the learning and performance of students in these areas. Student performance on high stakes exams has been used to measure program success across the program's development. Using data from different comparison groups, including same school historical data, same teacher historical data, same school data in classrooms not using the PERC model, and city-wide data, PERC students have passed state LE exams at consistently higher rates than their non-PERC counterparts in four schools over multiple teachers. Results from field trials (2009–2011) show strong performance compared to the available, most relevant comparison groups, averaging 17% points difference in LE exam passing rates. PERC-LE has also been implemented in summer school for students with histories of failure. In the 2011 summer school, 83% of PERC-LE students passed LE examinations, compared to a city-wide passing rate of 29% among students not experiencing PERC.

Increased data sharing among partners has enabled student-level analysis of student performance. Data from academic year 2011–2012 were analyzed using propensity score matching,[8] demonstrating significant differences between students in PERC classes and matched students in peer schools. Based on preliminary analyses, students in PERC-LE classes were 2.05 times more likely to pass the Regents exam than their matched peers. They were 1.51 times more likely to score at a level considered college-ready on this exam than their matched peers. The 2011–2012 cohort of TAS also performed significantly better on high stakes exams than their matched peers. Further analysis showed them to be 2.38 times more likely to score at a college-ready level in mathematics and science than their matched peers.[9]

Thus, the PERC model has demonstrated an ability to improve equity and access to learning for underserved students in urban schools. This chapter analyzes the model and the ways it substantially alters interactions among students and with their teachers, resulting in significantly improved outcomes for students. The research questions that we explore here are: (1) In what ways do student and teacher roles and responsibilities change when the PERC program is implemented? (2) How do interactions among students and between students and teachers in the PERC program compare to traditional settings?

Theoretical Framework and Literature Review

The PERC model is based upon the premise that students learn effectively when they work together under the guidance of a trained peer leader. Further, being a peer leader benefits those students in content understanding, academic skills, and self-concept. Being a peer leader requires training in content and pedagogy in a separate course, as well as management and monitoring by the teacher during the PERC class. The model is built upon a strong literature base concerning collaborative learning, peer tutoring/peer instruction, and formative assessment.

The PERC model employs peer-assisted team learning techniques that share characteristics of small-group cooperative learning practices.[10] However, PERC learning teams move beyond past concepts of student collaboration in both their daily implementation as the central instructional approach and in the leadership of a slightly older, trained peer instructor, who is not an honors student, the TAS. Past studies show that students who receive tutoring and peer-instruction receive widespread benefits, including improvements in academic performance, critical thinking, behavior, attendance, social skills, and self-confidence.[11] Some authors[12] have proposed that the recent, common experiences of *near-peers* are a reason for these positive impacts of peer tutoring. English Language Learners (ELLs) in particular have demonstrated gains in content knowledge and a range of academic skills through structured peer interactions.[13] In a review of the literature, positive impacts of peer tutoring on secondary students with mild disabilities were found, especially if the peer tutors were trained and monitored by the teacher.[14] The PERC classroom builds upon this literature, ensuring that students benefit from peer instruction every day for the majority of their class time.

In addition to the documented benefits to students being tutored, research has demonstrated a range of benefits to students who act as peer leaders or tutors.[15] Eighty-seven percent of studies in a review of the mathematics peer tutoring literature reported academic improvements among tutors.[16] Several of these studies found an increase in self-concept among peer tutors, and that tutoring has positive impacts for a range of ethnic groups.

A dramatic difference between the structures of PERC and traditional classes is the amount of feedback that students receive. TAS class includes extensive training in formative assessment techniques. Research on formative assessment has shaped this component of the TAS class to include analysis of learning expectations,[17] use of questioning and informal assessments,[18] and feedback to students.[19] Reviews of the literature and meta-analyses have demonstrated substantial positive impacts of formative assessment.[20] Further, increased benefits in learning by tutors when they engage in questioning as opposed to explaining behavior have been demonstrated.[21]

Methods

This study uses qualitative methods to explore and describe the program and its impacts, exploring and illuminating explanations for the program's success.

Participants

Participants include students in PERC-LE classes, TAS for the PERC-LE classes, and PERC-LE teachers. The PERC model has been developed within a public school system that serves over one million students, with 145,000 were designated as English Language Learners.[22] The four-year graduation rate in 2012 was 64.7% overall, with less than 17% of students earning an Advanced Regents diploma, which is a proxy for college-readiness.[23] More than 68% of these public school students meet or exceed the criteria for free and reduced lunch.

The four-year graduation rates in New York City[3] illustrate a significant opportunity gap. In 2012, 55.9% of Black students and 54.0% of Latinx students graduated in four years as compared with 75.0% of White students. Only 6.6% of Black students and 8.5% of Latinx students earned the Advanced Regents diploma, compared to 31.2% of White students. Students designated as ELLs had a 2012 four-year graduation rate of 36.2%, with 6.8% earning an Advanced Regents Diploma, compared to 64.8% of English Proficient students, 17.9% of whom earned an Advanced Regents Diploma. Closing this gap is a high priority within the city Department of Education, and for the PERC program. The population served by the program is representative of students in district non-selective high schools. In the district as a whole, only 1% of participants are White, one third speak a language other than English at home, and 30–40% of students entering the high schools meet or exceed 8th grade standards. For the 2011–2012 academic year, less than 2% of PERC students were White, 50% were Black, and 41% were Latinx, with 30% having a language other than English as their native language, and 67% qualifying for free or reduced price lunch. On 8th grade exams, 26% scored proficient in math and 17% in English. Among the TAS, less than 2% were White, 45% were Black, and 40% were Latinx, with 19% having a language

---

3 Editorial Note: Given the rapid pace of data capture and evolving data uses, data reported may differ depending on when the end-user accesses online resources. Parameters for the publication of archived and historic reports or documents that contain information, including statistics, are determined by individual entities at various levels of organization and accountability. The call for data transparency is evolving and represents a critical means for examining education statistics.

other than English as their native language, and 72% qualifying for free or reduced price lunch. On 8th grade exams, 48% of TAS scored proficient in math and 37% in English. Eight Living Environment teachers participated in the program from 2009–2012, the formative years of the focus on TAS within the program. The teachers' professional experience at PERC program entry ranged from one to 15 years. Most teachers were female and White, with one African American and one Latinx teacher.

Data Sources

Data describing participants' interactions have been collected through interviews, focus groups, surveys, and reflective writing across the years that the model has been developed. Teachers, TAS, PERC students, and program staff regularly share their experiences and perspectives as part of their program participation. Interviews are recorded and transcribed. Classroom observations are recorded through field notes and observation protocols, which are informed by the PERC target behaviors matrix (Appendix I), which was developed and revised during this study. In order to explore the two research questions, we reviewed the transcripts, survey responses and reflective writings, culling the responses for references to interactions among students or between students and teachers. We also conducted interviews of current PERC teachers and professional development staff and surveys of current TAS about these questions.

Findings and Analysis

The various qualitative data sets were analyzed and interpreted to develop findings, themes, and suggestions.[24] Thorough reiterative readings, patterns emerged from the data sets that illuminated the relationships in the classrooms.[25] We obtained participant feedback on our findings through a research presentation to PERC teachers and their TAS. We used their feedback to confirm and refine our analysis.

The Peer Enabled Restructured Classroom (PERC) Model

First piloted in a partner school in the spring of 2008, the PERC model has been developed, tested and refined to determine the critical elements for effective implementation. While the program recognizes the need of individual teachers and schools to adapt the model to specific instructional contexts and that individual lessons will vary to some extent within the parameters of the model, the teachers, professional development team, and researchers involved in the project have developed a set of target behaviors for the students, TAS, and teachers in PERC classrooms (see Appendix I) during this study.

In PERC-LE classrooms there are about 30 9th grade LE students and seven 10th grade TAS. In PERC-LE, the teacher begins class by providing a conceptual context and preliminary instruction, then at least 30 minutes of each class each day is spent in small group teamwork led by the TAS. In a well-implemented PERC class, students stay on task and discuss biology, guided by TAS. They follow instructions provided by TAS, ask TAS questions when they are confused about content or task requirements, answer scaffolding questions posed by TAS, and have TAS check the quality of their work before moving on to a new task. TAS keep students on task, ensuring that students understand both the content and the task requirements. They monitor student progress and conduct formative assessments. During group work time, the teacher moves among groups, listening to discussion and assessing understanding. The teacher contributes to TAS group discussions to model effective questioning for the TAS and push the group to higher achievement. The PERC teachers also use this time to work with individual students who need remediation or enrichment. At the end of class, having discussed their ideas in small groups, PERC students contribute to whole class discussions and answer questions posed by the teacher during whole class summaries.

The experience of the TAS is intended to reinforce and enrich conceptual understandings in the content areas for the TAS, and address 21st century skills such as goal-setting, monitoring, and questioning.[26] Each TAS participates daily in a special class led by the PERC classroom teacher. In this class, TAS practice teaching and assessment techniques, review PERC upcoming classroom lessons, give feedback to the teacher about the PERC students, and reflect upon and extend prior learning. The TAS then use these skills to shape interactions with their students in PERC classes. Additionally, TAS work on tasks specifically designed for them to develop college-ready skills during their TAS class.

PERC Student–TAS Interactions

Observational records, interviews, focus groups, and surveys with both teachers and students indicate that student-student dynamics in PERC classes contrast markedly with student behaviors in non-PERC classes. The PERC teachers reported that student-student interactions in traditional classes are mostly social in nature, frequently involving teasing, with work-related interactions usually consisting of one student giving another student the answers to classwork or homework questions. As one teacher explained, "I don't think in other classes that the students give credibility to any other student." Another PERC teacher contrasted the student interactions in PERC and non-PERC classes: "I think that normal student-to-student relationships, what I recognize is, it's more 'here are the answers' and teasing. And that's how they interact with one another. Whereas the TAS student relationships are more of trying to really keep the students on task and helping them with the assignment rather than giving them the answers."

While some TAS claimed to respect and help peers in other classes similar to their PERC students, most TAS described more differences than commonalities between PERC and traditional classes. PERC teachers claimed that the peer helping and peer respect demonstrated within TAS groups are unique within their schools. A TAS described the differences between interactions with her PERC students and students in other classes, "In the other classes, I have friends and we play around and talk but in my tutoring [PERC] class, I can't act this way because I have to be a role model for the students." In comparing his relationships, another TAS argued,

> My interactions with my PERC students, it's different compared to students in my other class. When I'm in my PERC class I'm always more mature and responsible than in my other classes so I can show them or guide to be mature and responsible.

Defining how PERC students think of their TAS, a TAS claimed, "A TAS is like a teacher that is closer to them than their actual teachers." Teachers argued that the TAS feel accountable for the success of their students and that TAS strategize about ways to support students who are under-performing. They contrasted this with students in other classes, "TAS just have a personal investment in other people that you don't always see. You actually don't usually see in general classes. They're not looking out for themselves. They're looking out for other people." According to these PERC participants, the TAS-student relationship is about academic success rather than social interactions.

Interactions within the TAS groups are the core component of the PERC model. TAS take on uncommon roles for students, which they identified as ensuring that their groups understand content, stay on task and complete their work, and are prepared for the state Regents exam. The TAS described themselves as leaders, guides, older siblings, and role models, labeling their relationship with their PERC students as "professional." As one TAS explained,

> My responsibilities and roles in PERC class is to make sure every one of my students understand what they are doing. My job is to help them understand about the concept and to look out for them. I have to make sure they are on top of their stuff and make sure they are completing their assignments.

Teachers reported PERC class benefits including elimination of classroom management problems, increased student time on task, and deepening of cognitive demands, all facilitated by the TAS. TAS claimed that the PERC class would remain the same if the teacher left the room, but that the class would completely alter in the absence of the TAS. This prediction is confirmed by teachers who describe chaos in their classes when their TAS are away on a fieldtrip but highly effective learning when a substitute replaces a teacher. One PERC principal tells about taking a state assessor into a PERC class and only realizing when the assessor wanted to congratulate the teacher for the in-depth learning happening in the TAS-led groups that a substitute was sitting quietly in the corner.

The TAS develop critical skills during TAS class that they apply with their PERC students. A vital piece of this is the formative assessment the TAS provide to the students, as the teachers underscore the need for the TAS to maintain "quality control" and make sure the students do not invest too much time in work that is incorrect. TAS explained that students are more comfortable asking a peer a question than showing ignorance to a teacher or in front of the whole class. The TAS emphasized the need to motivate their students, and the challenge that this poses for them. One TAS prioritized the "motivation the TAS provides. Staying on task is very important to keep students engaged and learning." TAS claimed that patience is one of the most important skills they develop as teachers, referring to both helping students understand scientific concepts and curbing their attempts to be off task. TAS work to balance the desire to be friends with their students and the need to be part of the instructional team.

Participants reported that being a role model for the PERC students is one of the most essential contributions of the TAS. A TAS explained, "[We] are role models because we try to be an example for the students. We try to show them the right way in how to act towards work." Another added, "TAS are role models because basically setting the example motivates other students to get engaged. The students will look at the TAS as though they're role models because they depend on us for guidance." Students, particularly ELL and former ELL students,[27] described the importance of interacting with someone like themselves who is academically successful. Teachers claimed that students have never seen scores as high as those earned by their TAS, who are not members of an elite honors group. Teachers believed that the students are motivated by this success of students who are like themselves. Additionally, teachers and TAS identified marked changes in behaviors by the TAS between their 9th grade year and the year when they serve as TAS. The TAS claimed that this role has positive impacts on themselves as students and their orientation toward academic success. For example, one TAS explained, "You can't cut classes when you are a role model."

Teachers work to get the TAS to see that they are role models throughout their school day, which is an ongoing effort.

PERC Teacher PERC Student Interactions

The PERC model requires role changes for teachers, shifting their primary responsibility from "being a dispenser of information to an assessor of understanding" and managing an instructional team. PERC teachers learn to access detailed information about their students by listening to TAS group discussions and from reports by their TAS. The teachers claimed that they do much more listening to students in PERC classes than in traditional classes. For example, one teacher argued, "rather than constantly telling them things the way I do to my students in my regular classes it's more me checking for their understanding in what they should have covered in their groups with their TAS and just reinforcing that." Additionally, because the TAS keep most students on task throughout the class, teachers reported that they can have in depth conversations with students at the extremes of performance that are not possible during traditional classes. One teacher described the contrast between PERC and other classes, "You really get to see the kids that need your help more," because the TAS are able to support the majority of the students, "So, then you just check on those students to make sure that they're good, and I had the free time to help the two that didn't understand. I would have never have been able to do that in a regular class." A TAS explained, "Teachers can single out students who really do not understand concepts and help them develop their understanding. The TAS can help the students who have a basic understanding and develop their understanding of the lesson." The teachers and TAS valued this ability to differentiate instruction and maximize teacher impact that is facilitated by the PERC class structure.

There was some variability among the teachers in the ways that they perceive their relationship with the PERC students. For example, some teachers initially reported that the TAS act as a buffer between themselves and the PERC students. However, when questioned about their understanding of past students, these teachers reflected that their assessments of prior students' understanding were largely based upon verbal responses by a few students during whole class instruction. They began to question how well they had known these students in their traditional classes and consider the diversity actually present in their classes. Some teachers who turned grading classwork and homework over to the TAS completely also reported that they did not feel able to monitor student progress. Encouraged by the professional development team to change these practices, these teachers felt better informed once they began regularly reviewing the work graded by the TAS.

Most teachers agreed that the TAS know the students better than the teachers know them, but the majority of teachers viewed this as an asset, arguing that they could never access the kind of information that students reveal to each other. One PERC teacher explained, "The TAS help the teacher know more details about the students, such as strengths, misconceptions, weakness. The TAS often know issues from the students' personal life." For example, teachers found that TAS have important insights about group composition:

> They know who works well with who else. They know who should be separated, and I feel that them having input and us working together to create the groups was really helpful, and we really were able to analyze those different personalities and abilities and come up with some really great groups now.

A TAS explained, "the TAS understand their students more because we knew at one point in time that that was us, we had that same misunderstanding at one point." Another TAS added, "We are around their age and basically we understand the thing that they don't and the reasons and difficulties they may experience." While these TAS highlighted their ability to understand student misconceptions, teachers commented that the TAS frequently have higher expectations for their students than the teachers do. During a class observation, one PERC teacher remarked to her coach, "I never would have asked that student a question that difficult, but he answered it." The teachers who used the TAS as sources of information about the students while maintaining direct contact with student work seemed to be the most comfortable with the triad dynamic. As a TAS articulated, "The TAS also are like supporting ladders that are set up to help the teacher to help the student to get to their goal, so the TAS are the closer to students than teachers in a regular class."

PERC Teacher–TAS Interactions

The Teacher-TAS relationship is unique for most program participants. TAS reported treasuring relationships that they develop with their PERC teachers, feeling that teachers treated them as vital team members and valued their input into instructional decisions. As one described, "with the title of being a TAS, teachers tend to look at us more of a pair than normal students. As a TAS, I feel more respected by other teachers too." Another TAS reflected, "PERC allows teachers and students to look at each other in a different way and get to know each other on a higher level." TAS claimed that these interactions boosted their confidence and increased their investment in school. For example, "Being a TAS helped me grow as a student and a person. It made me want to become a teacher even more than I wanted to before being a TAS." While some TAS alleged that their interactions with their PERC teacher were not unusual because they treated all teachers with respect, most TAS indicated that working with the PERC teachers across multiple years and during two periods a day strengthened their relationship. One TAS explained, "I feel that the PERC teacher's relationship with us TAS is also influencing the TAS life because after working multiple years with the teachers then you feel free with them and they help push you into being a better person." They attributed part of this closeness to these teachers to shared conversations about student work and planning PERC classes. Teachers also recognized the important role that this collaboration plays in forming their unique relationship with the TAS. One teacher described the way her interactions with TAS differ from regular students,

> I'll ask the TAS what they think in terms of how to explain something or how a lesson should flow or I look to them for recommendations and things like that on what they think would improve the lesson or where they think their students would have a hard time or where they have had a hard time and what we need to clarify those concepts. And they're almost like coworkers in a way more so than just students.

Teachers described their esteem for the TAS, the high standards to which they hold the TAS, and the fact that the vast majority of TAS live up to those expectations, highlighting the importance and complexity of their relationship with the TAS. They described their first observations of TAS as eye opening about the potential of students to teach each other and to perform well themselves. This first observation happens during spring classroom visits during new teacher orientation or during summer school professional development, both with TAS who have been trained by PERC teachers over many months. The TAS make such progress over the academic year that teachers are amazed each fall at how much work it takes to develop both the dispositions and skills necessary to be an effective TAS. New PERC teachers expressed feeling challenged by learning to trust the TAS and share control of the PERC classroom with them, but also explained dramatic successes once they learned to do this. One PERC teacher agreed, "in order for the PERC model to work you have to trust your TAS are going to properly support your students, which is really difficult at first." While the teachers learn to treat the TAS as educational partners, they also emphasize the need to remember that the TAS are students themselves who make mistakes and need support.

Implications

The PERC model forces students and teachers to take on new roles and responsibilities in the science classroom, transforming the relationships among all participants. PERC teachers learn to share the leadership of the learning experience with average adolescents who grow into role models and instructional guides for their peers. The teachers develop trust for the TAS and help them fulfill their potential as college-ready students. These respectful relationships help change the way that the TAS think about themselves and the goals that they set for their academic performance and classroom behavior. PERC students benefit from seeing that students just like themselves can have an academic focus and succeed on high stakes assessments. In participating schools, student-student interactions in traditional classes rarely support student learning. In PERC classes, students feel responsible for each other's achievement, working to keep each other focused on the task and helping each other understand content and skills. In PERC classes, teachers have the opportunity to listen to students, developing vital insights into student understanding and true academic potential, and providing targeted support to students on the extremes of performance within the class. The teacher leads an instructional team that collaborates to facilitate student engagement and success.

This research furthers learning about how peer-facilitation of instruction and teacher support for peer-to-peer experiences help students improve competencies in STEM, and help teachers address the demands of urban education. PERC-LE is a promising approach to improving achievement in science among historically under-performing groups in urban non-selective high schools. The intervention has two important implications for participating schools: (1) strong achievement benefits at two levels are evidenced, for students in the 9th grade level learning teams, and for 10th grade TAS team-leaders; (2) because PERC-LE uses students as teaching assistants (assets), it is sustainable at low cost to schools, with expenses only in the first few years of teacher professional development. Further, the relationships that PERC-LE develops within the 9th grade science classroom have the potential to transform teacher-student and student-student interactions across the school. The partnership is now working with teachers in advanced mathematics and science courses to create student-centered classrooms that build upon the collaborative skills and dispositions that students and TAS developed in the PERC classroom. A TAS-Pipeline-to-College is being created to support the TAS's continued growth and development through the rest of high school and into college. The partnership is exploring the long-term impacts of participating in the PERC classroom and continuing through the pipeline, comparing performance of participants to matched students in non-PERC schools and exploring the reasons for program success and needs for modification.

Discussion Questions

- What are some ways you can increase the use of collaborative learning and peer instruction in your context?

- How can you facilitate questioning rather than explaining behaviors by teachers and students?
- What kinds of authentic leadership structures you can co-create with learners?
- What near peer role models do learners have regular interactions with?
- In what ways might using a TAS address a teacher's need to control curriculum or cultivate student voice?
- What are the values and consequences of the *triad dynamic* in this type of classroom?
- How might a shift in student roles toward becoming a STEM TAS in your teaching or learning context impact the STEM teacher shortage in your community? What are the supports or barriers for such a shift?

Notes

1. This is the subject of research presented by Johnson, Schendel, McClellan Ribble, & Liu on institutional capacity building later in this volume.
2. Common Core State Standards Initiative, 2012; Conley, 2008; The White House, 2009.
3. New York City Department of Education, 2012.
4. Adelman, 2006; Porter, McMaken, Hwang, & Yang, 2011.
5. National Center for Education Statistics, 2012a, 2017.
6. *Ibid.*
7. *Ibid.*
8. Everson, Bonner, & Thomas, 2013.
9. *Ibid.*
10. Ginsburg-Block, Rohrbeck, & Fantuzzo, 2006.
11. Ginsburg-Block et al., 2006; McMaster, Fuchs & Fuchs, 2006; Morrison, 2004; Robinson, Schofield, & Steers-Wentzell, 2005; Shamir, Zion, & Ornit-Spector, 2008.
12. Lockspeiser, O'Sullivan, Teherani, & Muller, 2008.
13. Calderón, 2007; Chamot, 2009; Curtin, 2009; Faltis & Coulter, 2008; Fathman & Crowther, 2006; Meltzer & Hamann, 2005; Peregoy & Boyle, 2008.
14. Stenhoff & Lingaris/Kraft, 2007.
15. Ginsburg-Block et al., 2006; Robinson et al., 2005; Roscoe & Chi, 2008; Topping, 2005.
16. Robinson et al., 2005.
17. McMillan, 2003.
18. Ruiz-Primo & Furtak, 2007.
19. Brookhart, 2001.
20. Black & William, 1998; Kingston & Nash, 2011.
21. Roscoe & Chi, 2007.
22. NYC DOE, 2013a.
23. NYC DOE, 2013b.

24. Yin, 2009.
25. Huberman & Miles, 1994.
26. National Research Council, 2010; National Science Teachers Association, 2011.
27. Reported previously in Gerena & Keiler, 2012.

References

Adelman, C. (2006). *The toolbox revisited: Paths to degree completion from high school through college.* Washington, DC: U.S. Department of Education.

Black, P., & William, D. (1998). Assessment and classroom learning. *Assessment in Education: Principles, Policy & Practice, 5*(1), 7–75.

Everson H. T., Bonner, S. M., & Thomas, A. S. (2013). *Looking for early evidence of implementation: Using the lens of propensity score matching.* MSP Learning Network Conference. Washington, DC.

Brookhart, S. M. (2001). Successful students' formative and summative uses of assessment information. *Assessment in education: Principles, policy & practice, 8*(2), 153–169.

Calderón, M. (2007). *Teaching reading to English language learners, grades 6–12.* Thousand Oaks, CA: Corwin.

Chamot, A. U. (2009). *The CALLA handbook: Implementing the cognitive academic language learning approach* (2nd ed.). Boston, MA: Pearson Longman.

Common Core State Standards Initiative (2012). *Implementing the common core state standards.* Accessed February 21, 2013 at http://www.corestandards.org/

Conley, D. T. (2008). Rethinking college readiness. *New Directions for Higher Education, 2008*(144), 3–13.

Curtin, E. M. (2009). Teaching English language learners in the content areas: Language arts, social studies, math and science. In E. M. Curtin (Vol. Ed.) *Practical strategies for teaching English language learners* (pp. 141–174). Upper Saddle River, NJ: Pearson.

Faltis, C. J., & Coulter, C. A. (2008). Teaching and learning math for English learners. In C. J. Faltis & C. A. Coulter (Eds.), *Teaching English learners and immigrant students in secondary schools* (pp. 94–112). Upper Saddle River, NJ: Pearson/Merrill Prentice Hall.

Fathman, A. K., & Crowther, D. T. (2006). *Science for English language learners: K–12 classroom strategies.* Arlington, VA: NSTA Press.

Gerena, L. & Keiler, L. (2012). Effective intervention with urban secondary English Language Learners: How peer instructors support learning. *Bilingual Research Journal, 35*(1), 76–97.

Ginsburg-Block, M. D., Rohrbeck, C. A., & Fantuzzo, J. W. (2006). A meta-analytic review of social, self-concept, and behavioral outcomes of peer-assisted learning. *Journal of Educational Psychology, 98*(4), 732–749.

Huberman, A., & Miles, M. (1994). Data management and analysis. In N. Denzin & Y. Lincoln (Eds.). *Handbook of qualitative research* (pp. 428–444). London: Sage Publications.

Kingston, N., & Nash, B. (2011). Formative assessment: A meta-analysis and a call for research. *Educational Measurement: Issues and Practice, 30*(4), 28–37.

Lockspeiser, T. A., O'Sullivan, P., Teherani, A., & Muller, J. (2008). Understanding the experience of being taught by peers: The value of social and cognitive congruence. *Advances in Health Sciences Education, 13*(3), 361–372.

McMaster, K. L., Fuchs, D., & Fuchs, L. S. (2006). Research on peer-assisted learning strategies: The promise and limitations of peer-mediated instruction. *Reading & Writing Quarterly, 22*(1), 5–25.

McMillan, J. H. (2003). Understanding and improving teachers' classroom assessment decision making: Implications for theory and practice. *Educational Measurement: Issues and Practice, 22*(4), 34–43.

Meltzer, J., & Hamann, E. T. (2005). *Meeting the literacy development needs of adolescent English language learners through content-area learning: Part two: Focus on classroom teaching and learning strategies*. Providence, RI: The Education Alliance at Brown University. Retrieved from http://www.alliance.brown.edu/pubs/adlit/adell_litdv1.pdf

Morrison, M. (2004). Risk and responsibility: The potential of peer teaching to address negative leadership. *Improving Schools, 7*(3), 217–226.

National Center for Education Statistics (2012a). *The Nation's Report Card: Science 2011. National Assessment of Educational Progress at Grade 8* (Report No. NCES 2012–465). Washington, DC: Institute of Education Sciences; U.S. Department of Education, Retrieved from National Center for Education Statistics http://nces.ed.gov/nationsreportcard/pdf/main2011/2012465.pdf

National Center for Education Statistics. (2012b). *The Nation's Report Card: Science in action: Hands-on and interactive computer tasks from the 2009 Science Assessment* (Report No. NCES 2012–468). Washington, DC: Institute of Education Sciences; U.S. Department of Education. Retrieved from National Center for Education Statistics https://nces.ed.gov/nationsreportcard/pubs/main2009/2012468.aspx

National Center for Education Statistics. (2012c). *The Nation's Report Card: The Nation's Report Card: Science 2011 State Snapshot Report: New York State Grade 8 Public Schools*. Washington, DC: U.S. Department of Education, Institute of Education Sciences, National Center for Education Statistics, National Assessment of Educational Progress 2009 and 2011 Science Assessments (ERIC Document Reproduction Service No. ED532475)

National Center for Education Statistics. (2017). *The Nation's Report Card: The Nation's Report Card: 2015 Science Assessment*. Retrieved from https://www.nationsreportcard.gov/science_2015/#?grade=4

National Research Council. (2010). *Exploring the intersection of science education and 21st century skills: A workshop summary*. Washington, DC: The National Academies Press.

National Science Teachers Association (2011). *Quality science education and 21st-century skills*. Retrieved from http://www.nsta.org/about/positions/21stcentury.aspx

New York City Department of Education. (2013a). *About us: Statistical summary*. Retrieved from http://schools.nyc.gov/AboutUs/data/stats/default.htm

New York City Department of Education. (2013b). *Cohorts of 2001 through 2008 (Classes of 2005 through 2012) graduation outcomes*. Retrieved from http://schools.nyc.gov/Accountability/data/GraduationDropoutReports/default.htm

New York City Department of Education. (2012). *Educator guide: The New York City progress report high school 2011–12*. Retrieved January 3, 2013 from http://schools.nyc.gov/NR/rdonlyres/E25F8B70-1C47-4212-9D01-94EC0C56993C/0/EducatorGuide_HS_2012_10_23.pdf

Peregoy, S. F., & Boyle, O. F. (2008). *Reading, writing and learning in ESL* (5th ed.). Boston, MA: Allyn & Bacon.

Porter, A., McMaken, J., Hwang, J., & Yang, R. (2011). Common core standards. *Educational Researcher, 40*(3), 103–116.

Robinson, D. R., Schofield, J. W., & Steers-Wentzell, K. L. (2005). Peer and cross-age tutoring in math: Outcomes and their design implications. *Educational Psychology Review, 17*(4), 327–362.

Roscoe, R. D., & Chi, M. T. H. (2008). Tutor learning: The role of explaining and responding to questions. *Instructional Science: An International Journal of the Learning Sciences, 36*(4), 321–350.

Roscoe, R. D., & Chi, M. T. H. (2007). Understanding tutor learning: Knowledge-building and knowledge-telling in peer tutors' explanations and questions. *Review of Educational Research, 77*(4), 534–574.

Ruiz-Primo, M. A., & Furtak, E. M. (2007). Exploring teachers' informal formative assessment practices and students' understanding in the context of scientific inquiry. *Journal of Research in Science Teaching, 44*(1), 57–84.

Shamir, A., Zion, M., & Ornit-Spector, L. (2008). Peer tutoring, metacognitive processes and multimedia problem-based learning: The effect of mediation training on critical thinking. *Journal of Science Education and Technology, 17*(4), 384–398.

Stenhoff, D. M. & Lingugaris/Kraft, B. (2007). A review of the effects of peer tutoring on students with mild disabilities in secondary settings. *Exceptional Children, 74*(1), 8–30.

The White House. (2009). Fact sheet: The race to the top. Retrieved February 21, 2012 from http://www.whitehouse.gov/the-press-office/fact-sheet-race-top.

Topping, K. J. (2005). Trends in peer learning. *Educational Psychology, 25*(6), 631–645.

Yin, R. K. (2009). *Case study research: Design and methods. Fourth edition*. Thousand Oaks, CA: SAGE Publications, Inc.

CHAPTER FIVE

Early Engagement in Research as a Tool for Broadening Science Participation

CASSIE XU,[1] ROBERT NEWTON,[2] MARGARET TURRIN,[3] AND SUSAN VINCENT[4]

This research was supported as part of a larger NSF funded study #HRD-1561637

Abstract

Despite significant investments at multiple levels in educational infrastructure, technology, and targeted engagement, a significant gap remains in the recruitment and retention of African-American, Latinx, and Native American students in sciences courses, majors and careers. This gap is evident at our own institution,

---

1 Cassie Xu, Senior Staff Associate and Director, Office of Education and Outreach
  Lamont-Doherty Earth Observatory/Earth Institute, Columbia University
  cassie@ei.columbia.edu
2 Robert Newton, Senior Research Scientist
  Lamont-Doherty Earth Observatory/Earth Institute, Columbia University
  bnewton@ldeo.columbia.edu
3 Margaret Turrin, Senior Staff Associate
  Lamont-Doherty Earth Observatory/Earth Institute, Columbia University
  mkt@ldeo.columbia.edu
4 Susan Vincent, Retired Teacher
  Young Women's Leadership School of East Harlem

the Lamont-Doherty Earth Observatory (LDEO) of Columbia University, which is one of North America's premier scientific campuses, where a painfully small number of students and faculty are from underrepresented communities. Over the past 14 years, an educational outreach effort, the Secondary School Field Research Program (SSFRP), has developed more-or-less organically at Lamont. The program has had significant success in bringing students from under-represented ethnic and economic groups to its campus for immersive field-science research and in recruiting them into science majors at four-year colleges. We're doing this through a simple but strategic combination of activities that create a multi-layered and flexible network of relationships in which students can establish supportive communities for learning and growth.

Editorial Reflections

I, Joy, was accepted into Cornell University's Minority Introduction to Engineering Program as a junior in high school. It provided Black and Brown students with a residential experience on campus for one week in the summer. We stayed in the dormitory and went to orientation classes in each of the engineering disciplines. That is all I remember, perhaps because I chose to study in a discipline outside of engineering, or perhaps because I don't really remember feeling like I was a part of the landscape. Programs that bring youth on to campuses are challenging for many reasons, one of them being because the students will often disrupt the culture and traditions of that space. There is one other thing I remember, a girl from Cleveland named Tanya. I wonder what ever happened to her?

This article is important because the authors are making transparent some of these issues. They are not posing solutions but at least putting the issues out front for contemplation. I find that the "hallowed" corridors of elite institutions are this way—they know that sustainable programs are tied to outreach but outreach is uncomfortable.

One of the local high schools I, Janelle, have been lucky enough to partner with utilizes problem-based learning (PBLs) in every content area. I have served as a panelist for many of the student presentations, since they always need to present their final projects to outsiders with some degree of expertise in the topic of the students' work. I have been both a learner and a mentor in the efforts of this school over the past several years. Seeing what was possible in terms of transdisciplinary problem solving shaped my idea of what STEM really meant, far beyond its component letters. I watched a psychology class present their research on the gendered nature of toys. A biomedical class shared their designs for prosthetic limbs. And a language arts class addressed literacy rates in their own community. To me, all these projects represent STEM—authentic, relevant, engaging problem solving. All of the projects integrated art and creativity, and some degree of entrepreneurship. Similar to the experience of some of the participants in the upcoming chapter, students conclude their projects by presenting their research publicly; this is a promising practice for helping students develop STEM identity, self-efficacy, and public speaking skills.

One of my critiques of the PBLs I saw at the school was the students' lack of understanding of the scale of the problems they were tackling. They were presenting "solutions" to huge issues like illiteracy and homelessness. I think it is an important part of the learning process to help the students realize that they will not solve these problems with their projects, and to frame it that way is misleading. Language matters. Even if they slightly modified their titles from "ending homelessness" to "addressing homelessness," or from "preventing illiteracy" to "improving community literacy," it would help students better grasp the scale of the problems they are learning about. In terms of an Aspirations framework, it would help them cultivate a more authentic sense of agency so that they can apply their plans of action with realistic goals in mind. Similarly, STEM providers who are invested in equity also need to be mindful of our language in the ways we describe our work. While the authors of the upcoming chapter hope to contribute to diversification of the country's workforce, they simultaneously describe the lack of attention to diversity or inclusion in their program design.

While I offered feedback to the students as a panelist for these PBLs, I was also asked to provide the STEM coordinators and teachers with feedback. As terrific as the projects were, as in any work we do in schools, and certainly when utilizing equity metrics, there is room for improvement. Thinking of the work of Lisa Delpit,[1] it is important to help illuminate some of the gatekeeping mechanisms of power structures so that students can successfully navigate them if they choose to, but we have to begin by respecting and honoring the students. Socially just STEM education requires that we problematize our norms and assumptions. For example, place-based instruction is rooted in Indigenous epistemologies that de-center Western paradigms;[2] they do not simply occur because they are held at a physical site. The underrepresentation of people of color in STEM and STEM education is the effect of systemic racism not only in the past, but in our contemporary day to day lives. Equity metrics remind us that when we look at the "gap," rather than looking at what is "wrong with them," we need to look very critically at what is wrong with us, and by "us" I mean those of us who are already well embedded in institutions as gatekeepers.

Background

The Secondary School Field Research Program (SSFRP) is a six-week research experience—a full-time and project-based summer immersion program in earth and environmental science research. Our principal field site and outdoor classroom is Piermont Marsh, a component of the National Estuarine Research Reserve that extends over about 1,000 acres along the western bank of the Hudson River estuary in Rockland County. Most (approximately 80%) of participants come from public schools in New York City. While many of their peers ride the subway to summer jobs or internships in offices and stores, SSFRP students arrive at Columbia University's Morningside campus to catch a shuttle that brings them to the Lamont campus in southern Rockland County.

The program, which started as a small educational extension to a research grant, now serves approximately 55 high school students, 15 undergraduates and 10 high school science teachers each year. The program grew out of a chance encounter in early 2004. Co-founder Dr. Robert Newton met a group of middle school students while riding the subway who overheard him and a friend arguing about science. Newton and his friend wound up answering random science questions and joking around with the kids, which led to an invitation for a group of students to intern in Newton's lab. At the same time, co-founder Susan Vincent, a now-retired science teacher from The Young Women's Leadership School of East Harlem, was interning there through Columbia's Research Experiences for Teachers program. She was looking for a way to bring field experiences to her students and experimented with taking Newton's students to Piermont Marsh that summer in the SSFRP.

Unlike most secondary school research opportunities at universities, we have not tried to recruit students with histories of high academic achievement. We have explicitly eschewed the "low income/high achieving"[3] recruitment strategy, asking our partnering schools instead to send us students who will make good use of the program. "Our participant profile is: interest in science, likes to work in groups, and doesn't mind getting dirty." Another choice made by the co-founders from the beginning was to pay the student participants a reasonable stipend, without which it would be impossible to maintain a program serving students from groups currently under-represented in STEM careers. The results so far have far exceeded our expectations and, we believe, may provide one useful model for diversification of the country's scientific workforce.

Currently, the central scientific focus of the program is a long-term ecological study of a brackish marsh along the southwestern bank of Tappan Zee, the widest stretch of the Hudson River estuary, about 10 miles north of northern Manhattan. Sediment cores indicate that a wetland ecology has persisted in Piermont Marsh since shortly after the retreat of the Laurentide ice sheet at the beginning of the Holocene epoch. Throughout most of its history Piermont has been home to a diverse range of sedges and grasses, shifting to more drought indicative representatives during dry periods. During the 19th and early 20th centuries, salt hay grass (Spartina *patens*) was cultivated in meadows at the northern end for use by residents of the Village of Piermont. Today, however, the Marsh is dominated by an invasive reed, Phragmites *australis* (Phrag). At the beginning of the SSFRP, the program's research was on the displacement of native plant species by *Phragmites* and the effect of this invasive plant on habitat availability for fish. Research teams have sought to quantify the spread or encroachment of Phragmites in the marsh, understand the role of allelopathy in sub-soil competition, measure soil nutrients and root depth, and most recently, study non-chemical methods for managing this invasive species.

Over the last fourteen years, the research scope has broadened considerably. In the summer of 2017, students chose to work on one of twelve teams: micro-plastics in faunal tissue, micro-meteorites in core samples, nutrients in creeks and pore waters, sediment accumulation and characterization (including heavy metal content), lead content in marsh and urban soils, non-chemical *Phragmites* management techniques, fiddler crab census, methanogenesis, carbon sequestration in marsh soils, macroinvertebrate census, data analysis, and video documentation of the summer. In contrast to other young investigator programs where students are encouraged to design their own experiments, the scientific topics are determined by LDEO scientists, in consultation with scientists from New York State's Department of Environmental Conservation (NYS DEC), and the Palisades Interstate Park Commission (PIPC). The Marsh is owned by the PIPC (it abuts PIPC's Tallman State Park) and is managed by the NYS DEC on behalf of the National Oceanic and Atmospheric Administration (NOAA) through the Hudson River National Estuarine Research Reserve (HR-NERR) program. Our program design is closer to graduate student education, where students begin by working in a scientist's lab on existing projects and move on to their own research only after significant familiarity with the field. Some research (e.g., methanogenisis, lead in soils, and carbon storage) are aspects of ongoing funded research by Columbia investigators, and some are at the recommendation of the HR-NERR program (e.g., fiddler crabs, sediment accumulation). Others (e.g., fish counts, Marsh nutrient fluxes, micro-plastics in marine life) were established by the Program itself. Typically, the projects are multi-year, long-term observational studies. Students "plug in" to ongoing projects, inheriting the work of earlier participants and adding their own contributions. In addition to the Marsh work, students sometimes are placed in Lamont laboratory projects. For example, students have worked in labs that measure cosmogenic nuclides, noble gas isotopes, rare earths in minerals, and micro-fossils in sediment cores.

The program has always operated as a partnership between Columbia scientists and NYC public school teachers, originally between its co-founders, Newton and Vincent. More recently, the group has grown to engage approximately eight Lamont researchers, who make up its Steering Committee, and a network of about ten science teachers, all but one from New York City public high schools. Once the program was up and running in 2005, students and teachers were added from The Young Women's Leadership School (TYWLS) of East Harlem, the Urban Assembly New York Harbor School (Harbor School), the Frederick Douglass Academy I (FDA-I) and Curtis High School. Other schools, including 17 high schools associated with the Charles Hayden Foundation and several schools in Rockland and Bergen Counties are more recent additions. The cohort of partnering teachers has expanded more slowly, with a core from the original schools, and several recent additions.

The project had initial funding from a series of small private foundation grants.[4] Between 2008 and 2015, additional funding came from three National Science Foundation (NSF) education grants. While most NSF funding was directed to either graduate student education or teacher professional development, each of our projects had a summer component that overlapped with the SSFRP, and in aggregate these programs were important in sustaining the program's activities. Currently, core funding comes from three private foundations, the Charles Hayden and Pinkerton Foundations and The Young Women's Leadership Network. The first two are focused on youth development strategies, across several venues, targeting economically disadvantaged youth in New York City. The Young Women's Leadership Network, acts as a Learning Services Organization for a network of public girls' schools in New York City and Philadelphia. In addition, the program leverages educational supplements or broader impacts components of Lamont's research grants, and other occasional federal grant income. Public and private foundation support has allowed the program to hire a part-time administrator to carry tasks (applications, tracking participants, stipends, purchasing) that had previously been performed by teachers and science staff. The program has been transitioning, with partial success so far, to a model in which schools or school networks would fund their own teachers' participation and some of the student stipend costs would be borne by schools or city agencies.

Working with classroom teachers has been foundational to the program, and they have played critical roles in creating appropriate experiences for the students. During the first few years of the program, teachers undertook much of the organizational and administrative management, including acting as leaders for the research teams. In addition, teachers usually understand the emotional, interpersonal and pedagogical issues that teens, and specifically teens from their own schools, are dealing with better than their Lamont partners do.

In the past five years, the program has hired some of its alumni to return from college to work as research team leaders. For example, the 2016 season planning was kicked off by a group of twelve young women meeting at the Young Women's Leadership School in East Harlem. Each was either in an undergraduate science program or preparing to enter a graduate program in the fall. The alumni cohort was ethnically diverse and represented a range of institutions including the State University of New York (SUNY), the City University of New York (CUNY) schools, elite research universities, and private liberal arts schools. The undergrads were given day-to-day responsibility for team organization and logistics. They also provided an important layer of near-peer mentoring between the scientists/teachers and the students.

With undergraduates handling day-to-day activities over the past three to four years, teachers have reassessed their involvement, and their choices have been diverse. Several have reduced their time-commitment, and some have shifted to being more hands-on in the field and the lab, essentially treating the summer as a science internship. A couple have assumed more responsibility, taking on major sections of the program management. For example, one teacher is now responsible for recruiting, screening, and planning teacher involvement; another is responsible for overall management of operational issues during the summer and for winter field trips. A few have effectively taken on serving as a front-line scientist with a science researcher available as needed (e.g.: a project on sediment accumulation and micro-plastics). Finally, a couple are taking a step back from the summer research projects and using the time to work on curriculum. In 2016, for example, two teachers developed a carbon cycle curriculum working closely with a Barnard College professor who is a leading researcher on carbon sequestration.

The program founders always argued that having teachers and students work on research teams together would have a positive impact on teacher practice. In part, we believed that it would help them reconnect with science as a process. But we also that deconstructing the hierarchy of the classroom would expand teachers' understanding of what was possible for their students. We want them to see students engaging with science in informal settings, using skill sets and accruing experiences that are not easily practiced, or perhaps not highly valued, in the classroom and school settings. As teachers have "voted with their feet" over the past several years, we see that their own goals for the summer program are quite diverse, and that while some align with our own vision, some disagree, with some teachers strenuously working to maintain the formal structures, distance, and authority of the school setting. These disagreements have been discussed, mostly informally, among the program leadership and the teachers, but have yet to be addressed in a conclusive, or even serious, way.

Most of the teachers come to the program with an interest in engaging their own students. If a teacher is partnering with us, we will always take at least a couple of students from their school. In addition, we believe that over time, engagement with the program has a positive impact on the teaching practices of these teachers as well as on the schools where they work. In the case of TYWLS East Harlem, we know from discussions with teachers that "The Marsh Program," as they reference it, is a component of school culture: girls who are interested in science careers know that the Marsh Program will provide experience, a college recommendation and an important addition to their resumes. Students know that "Marsh kids" are more likely to get into good schools and more likely to get full scholarships. Students are aware that it's a summer job that comes with a pretty cool social scene and some very fun activities. So, the program has come to occupy a positive position in the culture of the school.

However, teacher demographics do not reflect those of the student participants, as most of our teachers are White. Studies have examined the effects on student absenteeism and suspension when there are large racial and socioeconomic mismatches between teachers and their students, and modest improvements in performance are made when students are taught by teachers of the same ethnicity.[5] Reports have shown that racial diversity benefits every workforce, with teaching being no exception, and it's been made clear that teachers of color tend to provide more culturally relevant teaching and better understand the situations that students of color may face.[6] We have not attempted to address this mismatch. To accept the help of nearly every teacher who has wanted to work with us. Fortuitously, the development of our cadre of undergraduate team leaders has helped to provide ethnically congruent mentorship. They have been able to develop trusting relationships with our young students, which we believe is an important factor in the program's success. While this works on a small scale for our program, the lack of teacher diversity does represent a challenge for those working in the STEM equity space. American public schools are now "majority minority," and NYC public schools are about 85% minority students, but the teaching faculty nationally is about 80% White, and about 60% White in New York City.[7] In science at the high school level, the disparity is even greater, with Blacks and Latinx accounting for only about 7% of science teachers nationwide. This misrepresentation speaks to a major issue in teacher preparation and recruitment and needs to be central in any discussion on STEM equity.

We have come to see the undergrads as a critical constituency for the program. The undergraduates are closely connected to the high school students; they are personally familiar with some of the students or their families. They have a solid understanding of the struggles that the younger students are grappling with. In many instances it is the undergrads who identify which students are having difficulties, which students are facing unusual personal challenges that need to be addressed (or at least taken into account), or which students need a change but may be reluctant to ask. In addition to running team operations, the undergraduate near-peer mentors have to lead scientific protocols on a day-to-day basis, teaching their assigned high school mentees general scientific methodologies (habits of mind, dispositions and practices), which require background reading and individual mentoring from their own scientist-mentors. So far, "SSFRP team leader" has been the first STEM job for all of them.

The workshop ran for three hours, with many of the participants in tears as they contributed their own stories to their younger mentees. In 2016, we offered a workshop in managing student debt. In 2017 we had an unusual number of students who reported they were thinking of transferring out of their colleges, or dropping out, from a sense of alienation and vulnerability in their spring semesters. We organized a workshop for them to exchange experiences and thoughts about causes and solutions. Our experience has shown that often young adults need a forum in which they can feel comfortable and safe in expressing their feelings; the Program has become such a space for many of our alumni; we intend to build on such programming in the future. The data reported here is a result of our ongoing tracking of our alumni and an unpublished study performed by the guidance office at the Young Women's Leadership School of East Harlem. For the past three years the program has been part of the NSF-funded (EHR Grant #1561637). Staying in Science study, which includes approximately 20 science mentorship programs for high school students in the NYC area. Our internal tracking will be merged with the Staying in Science results when those become available.

Program Structure

As currently structured, the program starts with an after-school Introduction to Wetlands Science course on Columbia's Morningside Heights campus. Taking a class on the campus contributes to a student's sense of belonging at the University and reinforces the authenticity of the upcoming research experience. Most of our participating students have not had prior experience on any college campus, and when they think about college, most do not feel that Columbia University is within their reach. Most program participants are Black, Latinx, or South Asian, whereas the Columbia campuses (including Lamont) are dominated by White and East Asian faces. Most of our participants are from poor or working-class families, whereas the dominant culture on Columbia University campuses is decidedly professional and upper middle class. Given this, there is a natural tendency among our students to feel apart from Columbia; having a connection to the physical campus is important to helping them build the sense of belonging that they need as they prepare for research work. In the early years of the program, activity during the school year depended entirely on teachers volunteering their time. At TYWLS, students participated in after-school science clubs, Young Science Achievers, and the New York Science and Engineering Fair. Vincent took some students to the Geologic Society of America and Society of Wetland Scientists meetings. As the program grew and more options became available, we have been able to offer the students a more robust program during the school year.

The summer research program runs for six weeks beginning after the Fourth of July. The first two days are dedicated to orientation and lab safety training, followed by more safety training for issues specific to Lamont and to working from small boats and in a wetland. The subsequent five weeks are dedicated to field and laboratory research. The last few days are set aside for students to complete their research presentations, a mix of short PowerPoint talks and posters. A few students work in classical laboratory internships (i.e., a pair of high school students in a Lamont researcher's laboratory) as junior level technicians. For example, in 2017, two students worked in a marine sediments lab, preparing samples, picking through washed sediments for potential micro-meteorites, and then taking samples to a scanning electron microscope facility. However, most students work on long-term projects in Piermont Marsh where they are placed in larger research teams. A typical team might have four to six high school students, an undergraduate team leader and a scientist-mentor. Most science leads are Lamont Research Professors, but we currently have two that are science teachers and, in the past, have had several graduate students or post-docs lead teams.

For the first five weeks, Monday through Thursday are fieldwork days or laboratory and Friday the high school students are scheduled for activities that involve developmental or reflective group activities. Lunchtime mentors include any Lamont scientist including graduate students, post-docs, staff associates, research scientists, or professors. Most are early career scientists: graduate students or post-doctoral researchers, mostly in their mid-twenties to early thirties. We have not investigated the volunteers' motivations. It is our sense anecdotally that for the younger scientists, communication and outreach are a more important component of their narratives of themselves than for the older, more established Lamont scientists. It may also be that for the early career scientists there are fewer teaching opportunities; or simply that they are, in general, more open to novel experiences. We have seen that as the program has grown, and the lunchtime mentoring has become widespread, more members of the senior science staff have tried mentoring; it is, in fact, the one area of the program that now has significant participation from the tenure-track Columbia professors. Nearly all of the scientists who participate as mentors one year continue in subsequent years.

We do ask students about their experience with mentors, and the responses are varied. Most report that the lunchtime conversations are among the best they have with Lamont staff and scientists, but others say that they had a hard time connecting with their mentors. Those who report positively often note that the early career scientists are more approachable and are more like-minded than the mentor-scientists running their research projects. The fact that the younger mentors are closer to students in age also helps students more at ease. Mentors report that the most common questions from the program participants pertain to how one: gets into a college with good science programming, picks a major, gets through college, gets into graduate school and picks research topics. The program participants are very curious about the "nuts and bolts" of progression from a graduating high school senior to professional researcher. Early career scientists are, of course, much closer to those focal points and more emotionally engaged with them, than their older colleagues. Our instructions to mentors are relatively simple. They are told that this is the student's time, and to follow the mentee's lead about whether they want to use the time for help with their projects or to talk about career issues, or simply to get to know the mentor. If the student seems too shy to start a conversation, our instruction to the mentors is to use the time to get to know the mentee. We suggest they ask about the student's school, their family background, their living situation, where their family came from, how they wound up at Lamont, how they're thinking about the future. We encourage them to use this as an opportunity to "get to know someone who is probably from a very different background than your own." Anecdotally, some mentors report that there is not enough structure in the conversations, and we are adding some structured questions as options for these sessions

Thus, the mentoring model is multi-layered, with peer, near-peer, group, and Principal Investigator (PI)-student relationships interweaving alongside a weekly consult with a mentor scientist (typically early-career). Alongside this mentoring "ecology," there is modest scaffolding: we have occasional lectures on topics related to our teams' study areas and each week there are journal readings and student-led article discussions within the research teams. In some years there are improvisational theater classes to help students bond and to speak out in public. Students present weekly in our "share out" sessions, telling the larger group what was accomplished and what is planned for next week. But none of these activities is centered. They are, rather, considered ancillary to the main activity of the summer: the student's scientific research.

Enfolding the research are always the interpersonal relationships that constitute the social context for student learning. The students we are trying to reach and the environment around them are not independent of each other but rather reciprocal and interdependent.

Thus, as the program has grown, the leadership, including the science mentors and the Team Leaders, have increased our focus on the personal and interpersonal development of the participants and the team dynamics. For the students, these dimensions remain mainly in the subtext, becoming explicit mainly when problems arise. But for the mentors and team leaders, they are explicit and central topics.

During the last week of the summer, each team is required to put together a short abstract of their work (as might be submitted to a science meeting), a research poster (as might be presented at such a meeting) and a PowerPoint presentation their work and results. The program culminates with a research symposium to which the whole Lamont campus is invited. The students give 10-minute presentations followed by a poster session. Students whose projects have yielded useful data have the opportunity during the academic year to present their posters at professional science conferences such as the Geological Society of America or Society of Wetland Scientists' annual meetings. Such trips are expensive, since they involve long-distance travel, hotel stays and registration fees. Trips that involve conference attendance are difficult for students and teachers who need to take time off from school in order to attend. Nevertheless, these trips can be transformative for the participants. Students who attend such meetings typically solidify their learning and have the experience of being treated as scientists by strangers. Our students have been called out in plenary by science leaders as "the future of our field." In poster sessions the high school students are commonly mistaken for college science majors, or even graduate students. One of our teams won "best student poster" at a regional symposium. These experiences appear to have a profound impact on how students see themselves, and especially how they conceptualize their future selves. Teachers who chaperone the trips also report positive experiences. Seeing their students present, and seeing student research taken seriously by graduate students and science professionals, solidifies the teachers' respect for the research projects and expands their understanding of what is possible for their students. The meetings also give teachers an opportunity to reconnect with their own scientific interests. We think that it helps them situate themselves in a continuum of scientists, which includes their students, themselves professional and researchers.

Since our students come from all five boroughs and several suburbs, and our work site is in Rockland, transportation has always been a major challenge for the program. In 2015, for example, we mapped all participants' commutes. The range was from 20 minutes to two hours, with an average of about 70 minutes, just to get to the Columbia shuttle. The shuttle to Lamont adds an additional 40 minutes, so most students commute well over an hour each way to participate in the program. In the early years the program was small enough that students could just join Lamonters on the inter-campus shuttle. But as the program grew, this got to be a significant friction point, as overcrowding meant that each day some people were standing and occasionally one or two Lamont employees would have to wait an hour for the next bus. In addition, only one of our partnering schools made transit fare cards available for students over the summer, so most students had to pay $5 per day in commute costs, which for them is a significant outlay. (One student missed the first day of the summer because he was arrested jumping the turnstile to get to the shuttle!) One of our private foundation funders insisted on the program providing transit fare cards to all students; a need that in retrospect we should have recognized much earlier.

Student engagement with data is another challenge that has been, and continues to be, significant for the program. We have discovered that young investigators such as ours usually do much better with the physical work at field sites and in laboratories than they do with processing and interpreting data. For nearly all of our new participants, this is their first experience generating "real" data themselves; and only a very few have had any training in data handling and statistics. Most are quick to believe that whatever they've done with their data is wrong; they are often surprised if it turns out that their intuitions about their own data turn out to be right. In the face of these challenges, the most common affect among our participants is a painful shyness. To address these difficulties, we have added a data team during the summer and a data analysis course in the fall. The data team works alongside the projects, ensuring that they begin the summer with a projection of what data they will produce and an appropriate data collection form. As the summer proceeds, they help their teams create spreadsheets, quality control their data, record meta data, and make charts. The fall course is computer-based, with students learning basic MATLAB, and using Excel and MATLAB to explore a couple of environmental data sets.

Results and Discussion

In terms of student outcomes, the program has been surprisingly successful. Surveys completed in 2010 and 2014 indicated that 100% of our approximately 300 alumni have gone to college, nearly all of them into four-year universities. About 40% of the graduates are, or are interested in, majoring in science or engineering, and the SSFRP is becoming a well-trod pathway to STEM careers for under-represented populations in New York City. We are currently conducting another round of surveys to reach and connect with alumni by tapping into our existing network of alumni. In 2010 the guidance counselor at one of our partnering schools reported that,

> To me, the most striking result was not the college acceptances or the increase in financial aid ... though that was pretty impressive. The thing that hit me was that nearly 100% of the SSFRP kids took a semester abroad in college. For their peers who did not do the program, that figure was about 20%. (Personal communication)

The SSFRP alumni who pursue science in college have a wide range of trajectories after graduation. Some have been very successful in the program, yet struggle to fulfil their goals in college. Jessica (pseudonym) started with SSFRP when she was a junior in high school. She worked on several projects, including one that built a microbe-driven battery from marsh soils. She returned from college to work as a team leader, excelling in managing her team's lab work. She recounted that one of the biggest lessons she learned in returning to the program, is that "being a leader sometimes means stepping out of the leadership role to give space to your students to become the leaders themselves." She graduated from a SUNY college with a major in environmental science but that weren't with grades strong enough for graduate school. A second alum, who also returned from college to work for the program over the summers, graduated with a degree in biology from a top-ranked private research-one (R1) university. She went on to get a master's degree in biotechnology management and is currently working in science administration. A third student, also one of our team leaders, took a minor in environmental science, and a major in Asian studies. She is currently working her way up the managerial ranks at Apple stores. A fourth graduated with a B. S. in neuroscience. Her original intent was to either go to medical school or pursue a Ph.D. in neuroscience research. For the moment, she is struggling with whether to apply to graduate school or to look for a job outside her field.

It is not obvious to us what one should make of these diverse professional outcomes. On one hand, we should expect that the interests of one's teenage years should fade through college and that students will engage with new desires sparked by their transformative college experiences. And, additionally, there are a myriad of factors that contribute to career success, many of which are outside the scope of our program. Anecdotally, however, we have seen a significant number of students who strongly desired a career in science or engineering and who were successful within the context of the SSFRP research internships, yet who either couldn't work their way through a science major or completed their major but could not get into an appropriate graduate program. It is important to consider what the program could do better to improve these outcomes in the long run.

All of the teachers who have participated in the program, approximately 30 in total, have remained in teaching. With one exception, they have all remained in public schools, most of them in the NYC public school system, and nearly all of those in non-exam-entry schools. While our sample is not large, this contrasts starkly with the NYC school system as a whole, where turnover rates are about 20% per year.[8] The program invites most teachers to return indefinitely. Some teachers participated in the program for several years and moved on; several have made it a permanent part of their summer professional development activities, often working alongside students to develop research and content-specific skills. In summer 2016, two earth science teachers partnered with a Lamont scientist to create a curriculum unit on the carbon cycle and climate. After reading a New York Times article about a group of scientists working in Iceland to sequester carbon dioxide through mineralization, one of the teachers brought the article in to share with her team and colleagues and discovered that the lead scientist on the project was next door to the SSFRP Director's office. The teachers recruited him to consult with them over several weeks on a module for the Earth Science classes at TYWLS Queens. At an end-of-summer wrap-up meeting, several teachers asked that the program re-orient teachers' roles to allow for curriculum development. One teacher said, "What I love about the SSFRP is that I'm at work, with colleagues, but for once I'm not crushed by reporting and grading and bureaucratic tasks." Another said, "Now that the college students are running the teams, I don't know what we're supposed to do." The group (teachers and the SSFRP leadership) decided to have each teacher suggest an individual plan at the beginning of the summer for a project that might either lean towards research internship or towards curriculum development.

One significant outcome of the Program is its impact on the culture of LDEO. From an initial mentoring cohort of one, the SSFRP quickly expanded to include three or four Lamont research professors and one or two graduate students as summer mentors. In 2015, eight scientists and staff members organized as a Steering Committee for the program. The number of individual lunchtime mentors has grown to over sixty. The Steering Committee makes decisions about recruitment strategy, pedagogical programming and research plans. Recently, subgroups of the Steering Committee were awarded two National Science Foundation research grants based in large part on their work with the SSFRP. One is to study the impact of research- and project-based education on moral development among teens. The other is to create a partnership of schools, universities, government agencies and informal education organizations to spin up a new set of research immersion programs focused on broadening STEM engagement among groups currently under-represented in STEM. The program has also become an option available to all Lamont-Doherty principal investigators working on broader impacts components for their federal grants. Several scientists have reported getting particularly positive feedback from review panels on the novelty and authenticity of their work with the SSFRP, especially as regards engaging with demographic groups that are poorly represented in the earth and environmental sciences. For example, early career scientist and Steering Committee member Dr. Einat Lev received an NSF CAREER grant, a highly competitive five-year award designed to assist early career scientists on the development of their academic careers through a combination of research, teaching, and disseminating new knowledge. Lev's integration of her research on fluid dynamics of volcanoes with curriculum development and piloting through the SSFRP got especially high marks in review. The SSFRP has become an integral part of the Lamont campus. In the summers it literally changes the complexion of the campus, and it is widely supported and praised for its positive contributions.

This is not to say that the integration of over 60 teens, the vast majority of them poor or working class people of color, into a largely White, largely upper middle class, older and intensely work-focused community has occurred without struggle. Perhaps the most problematic friction point has been the shuttle from the Morningside campus to Lamont. Each summer there are complaints from Lamont scientists about crowding and noise on the bus. Our teenaged participants also complain about their interactions with adults on the bus. They feel that some Lamont staff are sarcastic, condescending, and/or hostile. It is clear from working with Lamont facilities and dialoguing with Lamonters about these friction points, that there is very broad support for the program. But it is also clear that there is a wide range of sensitivities to the racial history of America, and that many at Lamont do not see or understand how institutional structures operate in a way that might charge what they see as race-neutral comments. That is to say, in the context of the racial and economic barriers that our participants face, attitudes and actions that do not actively welcome them, and do not actively challenge discriminatory structures, leave our interns caught in objectively racist structures. Our participants perceive those impacts, whether we at Lamont empathize with their perceptions or not.

Lessons Learned

Each student's path through high school, including the SSFRP, and into college is individual and the experiences that help or hinder student attainment form a complex web that is neither compactly describable nor predictable. In fact, one of the pleasures of working in the SSFRP has been the opportunity to be a part of hundreds of individual children's lives, and to witness the diversity of personal struggles. This is especially true since the program grew organically from very small beginnings without too-firm a design or well-articulated set of methods. Thus, there have been many ad hoc additions to and subtractions from the program, most of which came about through extensive discussion among the program's leadership in response to problems or opportunities that arose "on the ground." At the program's inception and during its developmental stages our guiding principles were largely based on our experience as educators and scientists. Becoming familiar with the literature on pedagogical methods, including experiential, project-based, problem-based, student-centered, place-based, and so on, was largely ex post facto—more often than not as preparation for writing funding proposals.

We believe this ground-up, organic process has served the program well. The hiring of alums who are college science majors grew out of the intense financial need of one of our program graduates. Undergraduate leaders and near-peer mentoring (strategies supported in the literature) are additions that have become cornerstones in our approach. Team-oriented research projects were originally a product of our need to stretch scarce mentoring resources among as many students as practical. Now we see the team structure as a major strength, providing the emotional and interpersonal context for student engagement in both cognitive and social growth. We introduced a data analysis course in response to our students' struggles creating meaningful graphs for their capstone presentations. The literature is rife with studies demonstrating that quantitative skills represent a core deficit among underrepresented minorities struggling to stay in science and engineering majors.

If we've picked our way towards increasingly useful programming it is because our guidepost has always been to center the needs of our students. Improving students' quantitative reasoning abilities may be the next frontier for the program. Data analysis, statistics, algebra and data visualization appear to be the stumbling blocks that are standing in the way of our alumni moving through their majors to science degrees and then on to graduate school. The introduction of a data team, beginning each team's summer with a discussion of the data expected to be produced by the end, and making the data analysis course mandatory for all participants have been first steps in that direction. We will surely discover some more over the next few years.

What Works

In the next section we discuss four aspects of the program that we believe can be generalized to other efforts to broaden participation in science education and scientific careers: student engagement through authentic research, a layered mentoring "ecology," team-oriented activities (academic and social) that create supportive learning communities and place-based learning.

1. Build strong student engagement, both intellectually and emotionally, with the practical work of science in the context of authentic research. Engagement, by which we mean that the learner is a physically, emotionally and cognitively active participant, is a critical factor in both cognitive development and interpersonal maturity.
2. Include multi-layered mentoring model that interweaves peer, near-peer, and scientist-student relationships. The program does not identify a specific developmental model that students are expected to fit into. Rather, it establishes an ecology of multiple mentoring relationships and situations. Each student finds her or his own path through this setting, meeting perceived needs and filling in gaps with help where it is available.
3. Establish team-based research and collaborative learning. Research teams allow students to feel part of a larger collective effort and understand that their role and work is important as part of a broader continuum. It also means that they are fulfilling their emotional and interpersonal needs within the context of a science learning project.
4. The importance of place is well-documented in the educational literature. The interesting thing for the SSFRP is that we realized that our students' learning is rooted in two novel places: the Piermont Marsh, where most of their field sites are located, and the Lamont-Doherty campus which, for our students, is at least as novel a setting as the Marsh.

Increased Engagement Through Communities of Practice Around Authentic Science Research

Engagement implies bilateral connection between the teacher/mentor and the student/mentee; the more holistic that dialogue, the deeper the engagement. A central principle of the SSFRP's practice is that authentic research, especially field work, facilitates close and impactful engagement. On the other hand, lack of authenticity (for example, when students are making repetitive measurements just to assimilate techniques or when the tools applied are incapable of measuring desired quantities to sufficient precision), subverts engagement. All participants, including both mentees and mentors, take the work less seriously and care less about the outcomes without authenticity. In the intimate interpersonal setting of a field team, there is no way to disguise such lack of sincerity. Conversely, when the research is authentic and the discoveries real and at least occasionally surprising, then mentors and mentees share the authentic excitement of real learning; or, in the case of failed experiments or disappointing observations, real disappointment and loss are also shared. In such situations the intensity of engagement shifts the students' vision off their own capacities and potentials. In sharing honestly-felt emotions with their mentors, students come to feel intuitively that they could stand in the mentor's shoes. They move that much closer to imagining futures in which they are professional scientists who might even mentor young people similar to their current selves.

In our culture, access to such transformative engagement with science is far from evenly distributed. Access requires entry to a research facility, a way to transport oneself to and from field or lab sites, the free time to volunteer on a consistent basis, parents who understand the value of the activity, a self-image compatible with dedicating valuable time to a science internship, the basic quantitative and verbal abilities (and cultural norms) to fit into a research setting. Even with such facilities, it is difficult to find a scientist willing to take on very young investigators. Without them, it is nearly impossible. And we would argue that it is exactly the lack of such opportunities that limits STEM recruitment from demographic groups that continue to be under-represented in scientific professions. This inequality in access helps to explain why people from underrepresented communities represent fewer than 10% of college graduates working in STEM.[9] LDEO scientific staff, including Columbia's Department of Earth and Environmental Sciences, highlights this startling gap in diversity well. There has been significant progress in hiring women—in about 15 years the department has gone from 4% to 25% female, for example. However, the Department of Earth and Environmental Sciences faculty and the Lamont research staff include only two non-White/non-Asian scientists.

Building a Community of Practice through Team-Based Learning

Teamwork in the SSFRP is both a practical solution to scarce human resources and a central philosophical approach. The co-founders, Vincent and Newton, are from very different backgrounds but they share a belief in grassroots collective action. As the program and its resources have expanded, it has allowed the leadership to balance the research teams to improve the development of communities of practice. Teams are now a mix of novices and veterans. Weaker undergraduate team leaders can be paired with more capable high school students. When there are interpersonal issues, students can be moved to new teams. A benefit of using alumni as part of the leadership team includes shared experience and empathy. For example, we had a student who had dug himself into such a hole with his team that he was asked to leave it. He wasn't hostile, but he couldn't connect consistently. He was often late, and would sometimes 'space out' in the field, losing track of team activities or wandering off. The program leadership discussed the problem extensively with the student, his team, and several of the more experienced team leaders. The leader of another team stepped in. "Don't worry; I got this." And she did, taking him on as both a project and a friend.

In retrospect, we might have expected the outcome. The team leader had, herself, been a problematic participant in her first year, struggling to find a place in the team structure of the program without fighting with her teammates. When, somewhat to our surprise, she asked to return, we paired her with another student who was similarly strong-willed, but who was having an easier time integrating with her teammates. By the end of the summer the two were close partners and stepped up to some of the programs most grueling tasks together.

Multi-Layered Mentoring

We believe that the range of mentoring relationships described above allows students to establish relationships that match their needs: it feeds back into deeper engagement and consolidates shifts in the students' vision of their future selves. These shifts include explicit program goals such a higher education, and moving towards STEM careers, as reflected in surveys, focus groups, and informal conversations with the students over the years. It also shows up, we suspect, in less tangible dimensions such as students' ability to express themselves to peers and mentors, their confidence in trying new things, and their pleasure in learning. Having a rich ecology of mentoring relationships is what allows us to forego a rigid pathway for students. The ability of each child to find her or his own pathway empowers the students, allowing them to be both more of an individual and more connected to their peers.

Place-Based Learning

Engagement with a physical place—place-based learning—grounds the learner in the local environment and can increase awareness and connectedness to their community.[10] For our students, the Lamont campus and the Marsh serve as dual places of learning. Both are novel, even exotic, for city kids from families without academic backgrounds. At the beginning of their summer experiences, there is always some hesitation and uncertainty about what these two unfamiliar places will bring. Arriving at the Lamont campus, which must feel both distant and very different from students' neighborhoods, can be overwhelming. New students are usually painfully quiet, looking for cues as to how to act. The program fills the void by having them focus right away on accomplishing the practical tasks of starting their projects: finding and organizing their field equipment, arranging transportation, ordering any missing lab supplies, marking and GPS-ing their sites, learning new protocols, making schedules, etc. These activities give the students purchase—a way to contribute and thereby to establish themselves as belonging. And, while the work gets done, Lamont, and for most students the Marsh, become the locus for the study of science as well as the community-building described above. As the students inhabit the campus and the marsh, both places have profound impacts on them. The Marsh is a physically demanding environment: the mud is many meters deep and no one working there escapes sinking up to their butt on a regular basis. Canoeing to and from the sites (which for most of our participants is a brand new mode of transport), moving equipment in and out (over a four-foot mud berm and through dense reed thickets twice their height), taking samples without contaminating them, and staying organized and following strict protocols while tired and covered with mud, are all challenges that are overcome in the context of intensely collaborative group experiences and under the pressure of having to generate publishable data. Very few of our participants have been in a wetland, and the novelty of the environment heightens the intensity of the experience; the Marsh thus forces participants to be present—to pay close attention. That grounding heightens emotional and interpersonal experiences and, it has appeared to us, solidifies the cognitive gains in ways that would be impossible in a classroom.

The Lamont campus offers a different, and along some dimensions, more challenging place to the SSFRP participants. Physically, the campus is, if anything, more inviting than the students' schools. There are hundreds of acres of lawn and woods, and the students have ample space (two labs of their own, a large "common" room, the Program Director's office, atrium spaces, the cafeteria and library) available to them. These are places where the students can spread out and work, talk or play together without the scrutiny that encumbers New York City school buildings. Students are encouraged to break from their work from time to time, take a hike in the woods and connect with the natural beauty that surrounds them. This environment stands in sharp contrast to the asphalt and concrete urban landscape in which they live. Interpersonally, on the other hand, the campus implicitly insists that the students adapt to a professional situation that is mostly foreign to them. They are expected to get quickly to work in the morning and to stay focused and on task all day. They have to approach Lamont scientists and technical staff as colleagues. They are expected to be careful custodians of their samples, their tools and their results. They are expected to take their work seriously. While there is plenty of relaxed play, there is little patience for irresponsibility. The team leaders, near-peer mentors, are especially circumspect of childish behavior.

While integrating new sets of skills, our participants are also dealing with the discordance of being placed into a largely White, nearly-entirely upper middle class, explicitly elite subculture. And while the teenagers adapt to Lamont, they are also having a major impact on the place. They (literally) darken the complexion of the summer population of the campus. They bring music and their own voices to the shuttle bus and to the outdoor areas where they work. They join Lamonters, on the basketball court and soccer pitch. They engage a significant fraction of the early career scientists at Friday lunches, and those experiences follow the "double discovery" model with both sides learning about a world in which they have little direct experience.

Final Thoughts

Open discussion with the full group of participants has been our approach to a number of issues related to diversity, including racial friction between the students and LDEO professionals, gender identity and gender-based tension among participants. The trigger for discussion has often been comments made to or near students on the shuttle bus. Students have been shy about raising such issues, and initially won't speak directly, partly out of concern for the reaction of the (mostly White, cis, hetero) adults in the room. As the conversations loosen up, the students have surprised the program leaders with their nuanced approach to the issues. The students are able to talk through those issues, argue their points, present evidence, and listen to each other, engaging with each other in very sophisticated ways. Overall, we aim to create a space where students can feel like they belong and can feel that they are welcomed. When they feel like they are part of a collective effort, and that their work is useful and they themselves are contributing to a larger endeavor, they can be themselves, try new ideas, and be able to fail on a task without feeling like failures as people. The program gives the students the emotional, interpersonal, cultural, and physical space that is needed for them to grow as individuals. Alongside this growth at a personal level, students are providing each other with a cultural well-being through comradery, where their different backgrounds and histories are valued.

Our participants have made, and continue to make, use of the SSFRP in a variety of ways, from enhancing cognitive skills to broadening their social circles to being in nature and much more. Our high school and college students in particular have leveraged the program into success against traditional academic metrics (entrance into college, receiving financial aid, and completing technical or scientific majors). While students' skills acquisition during their time with us has been considerable, the program leadership believes that student success is primarily a result of how more nuanced, personal and interpersonal characteristics have shifted for them. Such growth requires that the students be treated with care, respect, and love. We must make them feel welcomed and invited by allowing their cultures, interests, and history into our communities. We must reconnect them to their own strengths and gain their trust so that they can learn from us. We must respect them, so that they feel connected to us. Only then can all of us—students and mentors together—discover that science is for everyone.

Discussion questions

- How can research experiences expand teachers' understanding of student capacities and shift their thinking about creating STEM-engaged futures for their students?
- How can teacher participation, where they are learning alongside students, in research experiences enhance their own capacity and confidence levels for scientific field and lab research?
- What types of experiences support teachers' learning about science and engineering process?
- In informal learning spaces, such as museums and science learning centers, how has the community of practice model/approach to student engagement worked to expand students' capacities to further this kind of learning in other educational and learning settings?
- How does *early* engagement (i.e., involvement of students in science research *before* students attend college) contribute to positive and negative patterns of engagement?
- How does the intensity of engagement in informal learning experiences contribute to the development of a student's science identity?
- What are the unintended consequences of "broader impact" requirements for supporting systemic change in programming for equity?

Notes

1. Author of important transformative works such as *Other People's Children* and *Multiplication is for White People*.
2. Deloria & Wildcat, 2001.

3. Also known as "creaming" or "skimming"
4. Grants range from $2,500 to $20,000 from various public and private funding organizations.
5. Holt & Gershenson, 2015.
6. Partelow, Spong, Brown, & Johnson, 2017.
7. U.S. Department of Education, 2016.
8. McElroy, 2009.
9. Allen-Ramdial, & Campbell, 2014.
10. Hebert, & Lewandowski, 2017.

References

Allen-Ramdial, S. A., & Campbell, A. G. (2014). Reimagining the STEM pipeline: Advancing STEM diversity, persistence, and success. *Bioscience, 64*(7), 612–618.

Deloria, V., Jr. & Wildcat, D. R. (2001). *Power and place: Indian education in America.* Golden, CO: Fulcrum Publishing.

Hebert, T., & Lewandowski, J. L. (2017). Blending community and content through place-based science. *Collaborations: A Journal of Community-Based Research and Practice, 1*(1), 1–9.

Holt, S. B., & Gershenson, S. (2015). The Impact of teacher demographic representation on student attendance and suspension. *Institute of Labor Economics Discussion Papers,* No. 9554. Retrieved from http://ftp.iza.org/dp9554.pdf

McElroy, M. (2009, June 30). Science teacher retention: It takes more than just money. *American Association for the Advancement of Science.* Retrieved from https://www.aaas.org/news/science-teacher-retention-it-takes-more-money

Partelow, L., Spong, A., Brown, C., & Johnson, S. (2017). America needs more teachers of color and a more selective teaching profession. *Center for American Progress.* Retrieved from: https://www.americanprogress.org/issues/education-k-12/reports/2017/09/14/437667/america-needs-teachers-color-selective-teaching-profession/

U.S. Department of Education. (2016). The State of racial diversity in the educator workforce. Washington, DC: Policy and Program Studies Service; Office of Planning, Evaluation, & Policy Development. Retrieved from https://www2.ed.gov/rschstat/eval/highered/racial-diversity/state-racial-diversity-workforce.pdf

CHAPTER SIX

Promoting Middle School Students' Motivation and Persistence in an After-School Engineering Program

SRINJITA BHADURI,[1] ALEXANDRA GENDREAU,[2]
VARSHA SRIKANTH KOUSHIK,[3] TAMMY SUMNER,[4]
JOHN RISTVEY,[5] AND RANDY RUSSELL[6]

---

1 Srinjita Bhaduri, Doctoral Student
 University of Colorado—Boulder
 srinjita.bhaduri@colorado.edu
2 Alexandra Gendreau, Doctoral Student
 University of Colorado—Boulder
 alge6158@colorado.edu
3 Varsha Srikanth Koushik, Doctoral Student
 University of Colorado—Boulder
 vasr6678@colorado.edu
4 Tammy Sumner, Professor of Cognitive and Computer Science
 University of Colorado—Boulder
 sumner@colorado.edu
5 John Ristvey, Director
 University Corporation for Atmospheric Research (UCAR)
 jristvey@ucar.edu
6 Randy Russell, Educator/Instruction Designer
 University Corporation for Atmospheric Research (UCAR)
 rrussell@ucar.edu

Abstract

Engineering Experiences aims to understand and promote practices that develop students' motivations and capacities to pursue careers in STEM fields by providing afterschool engineering experiences in atmospheric and related sciences to middle school students. This collaborative project is working with low-income youth and is being implemented in partnership with a local chapter of the *I Have A Dream Foundation*. The program combines STEM learning with hands-on technology and engineering design activities to prepare students for success in high school STEM courses and the future STEM workforce. This article reports on a semester-long implementation of a curriculum on Unmanned Aerial Vehicles (UAV) also known as drones. Both qualitative and quantitative data on student engagement, persistence, and competencies in engineering design were collected, including surveys, short writing assignments, interviews and observations. Results indicate the participating youth found the UAV curriculum to be motivating and engaging, and their perspectives on engineering and the role of UAV in science and engineering were broadened and enhanced. However, participating youth had difficulties generalizing their engineering design skills to new design challenges.

Editorial Reflections

During my time in Wyoming, I, Joy, met some amazing youth. Not usually on campus but in the schools served by university outreach programs and grants. I got the chance to travel throughout the state from the northwestern corner near Utah down to the southeastern corner near Colorado and many places in between. I will never forget the Lady Eagles. National Science Foundation ITEST grants like the one described in the upcoming chapter bring technology education to schools as out-of-school time programs. One of the schools served by the grant that I supported was at a Reservation school. When I see the symbols of the Shoshone and Arapaho Nations I see geometry and pride. The symbols are everywhere. The Aspirations framework highlights how students transform their learning and develop agency through place-based education. That project focused on spatial reasoning and visualization but place was an important component of the analysis—once students learned to write codes for games and robotics, any observer could see how they used code to insert their own culture and landmarks into the curriculum.

Though the work described in the upcoming chapter was initially established as an after school engineering program, the team engaged in research throughout the project that examined what they needed to do differently to better meet the needs of the students.

The researchers realized they needed to connect their out-of-school program with the students' in-school experiences. The chapter therefore represents a STEM ecosystem, and the authors conclude with some important questions. My, Janelle's, story serves to add to the list of questions rather than answer them. In my work, I have encountered some truly outstanding STEM programs that are designed for low income students.[1] One of the best I have seen is the National Society of Black Engineers (NSBE) Summer Engineering Experience for Kids (SEEK). In several cities across the U.S., NSBE SEEK engages third, fourth, and fifth grade students in high need schools in an exciting and community-relevant hands-on engineering program for three weeks. I served as a judge for SEEK's week-end competitions multiple times, and was blown away. There are so many aspects of the program that align with the Equity Metric. The team leaders are near peer mentors drawn from colleges and universities across the U.S. and other countries; they are primarily people of color and they represent a spectrum of STEM majors. Each week's activities center around a specific theme, and the students develop skills and knowledge around that theme as they work in both small and whole groups. The program is dynamic—lots of movement, songs, chants, cheers, and competitions. It taps into students' creativity and innovation, and art is integrated throughout the activities. Parent support has been consistently high, and students' family members I heard from anecdotally were overwhelmingly pleased with the program, often describing a newly planted excitement for engineering and STEM pathways.

With all of these positive elements of an opportunity to learn program, one aspect of SEEK left me scratching my head. Where were the teachers? Where were the district STEM coaches? How would they see the leadership and excitement for learning the kids were demonstrating? How would they get ideas for community-relevant teaching methods that could potentially better engage their students? Who from the school or district was there to see the room full of families ranging from parents, grandparents, cousins and siblings during the day, clearly making this STEM learning experience a priority? For out-of-school educational experiences such as NSBE SEEK to realize their transformative potential on students' trajectories, school stakeholders should ideally connect with them. We do not have a system in place to intentionally and consistently build comprehensive STEM ecosystems. Tasked with the need to ask "what must we do differently?" this chapter offers all of us a challenge.

Introduction

According to the US Department of Education's National Center for Education Statistics,[2] over half (52%) of the students in US K-12 public schools were eligible for free and reduced lunch during 2013–2014, up from 38% in the 2000–2001 school year. The story in Colorado is much the same, with 42% of students eligible for free and reduced lunch, up from 27% during this same timeframe. With this change in our national demographic, what do we know about the needs of these young people in afterschool environments? What are their interests? How can they be provided with experiences commensurate with their peers who have greater opportunities? How can these young people achieve a brighter future through career paths in STEM? What are the unique aspirations of young people in middle school who will be in tomorrow's workforce? Researchers in Colorado set out to better understand these and other questions through a National Science Foundation (NSF) Innovative Technology Experiences for Students and Teachers (ITEST) project beginning in 2015.

The world of the middle school student, who range in age from 11–13, is one of extremes. On one extreme, younger middle school students closer to elementary age tend toward having energy and enthusiasm toward experiences that are interesting to them. On the other extreme, older middle schoolers closer to high school age are often less transparent in their interests and tend to be much more reserved as they interact with peers and their teachers. These extremes span social and economic boundaries including those from low-income families. In our present study, we set out to understand how students from low-income families participating in out-of-school time settings would respond to technology and engineering activities set in the context of atmospheric sciences. For the *Engineering Experiences* project, we partnered with a local chapter of the *I Have a Dream Foundation*, who have established afterschool programs reaching elementary, middle and high school students from low-income families. The *I Have a Dream Foundation* partnership brings several benefits to the *Engineering Experiences* program: they recruit students from our target demographic, they have established long-term mentoring relationships with their students, and they work directly in the school in which the students attend. We implemented a new engineering focused program within their existing structure to study how and under what conditions *Engineering Experiences* contributed to participating youths' motivation and career aspirations in STEM.

The pedagogical and content design of *Engineering Experiences* is based on research-based principles in out of school time (OST) learning from the U.S. Department of Education OST Framework. This framework was derived by an extensive review of prior research conducted by an expert panel; it identifies guidelines for the design and implementation of research-based OST programs.[3] The most relevant OST principles related to our curriculum are providing engaging learning experiences and aligning the OST program academically with the school day.

At the beginning of our design and development process for the OST curriculum, we tested some simple existing and new curricular materials within the *I*

Have a Dream program in order to learn about the Dreamers (students who participate in the program), the structure of the program, and how engineering materials might fit into the afterschool program. To accomplish this we worked with two different groups of students, having them explore balsa wood airplanes and the design, construction, and testing of paper sail cars. Because this was the very beginning of the project and we had not yet surveyed student interest, or developed curricula, the project developers chose the topics. Over the course of three weeks, we learned that there was a large variability in student attendance. Student participation remained inconsistent throughout the course of the three-week pilot. We were curious as to the reasons for inconsistency in student attendance. Post-surveys and interviews suggest that the projects used in the pilot—testing gliders and building sailcars—may not have been sufficiently challenging. Several students also voiced a strong desire to work directly with more "high tech" materials. In response to this feedback, we generated an extensive list of possible topics and projects and asked all youth to vote on their top preferences: students overwhelmingly expressed interest in working with unmanned aerial vehicles (UAV) or drones. Drones proved to be an excellent choice, providing both opportunities for hands-on activities to youth and a coherent structure for integrating a broad range of engineering phenomena such as load testing, remote sensing, engineering design, and tradeoff analyses. Over the course of several iterations, we found that overall the curriculum worked with the students who attended and actively participated in *Engineering Experiences*.

Relevant Research on Cultivating STEM Interest in Youth

In this project, we implemented variations of work done to address student interest[4] and utility value interventions.[5,6] We adapted these interventions to make them more accessible to a younger cohort of middle-school students, some of whom were non-native English speakers. We integrated the utility value interventions into the curriculum in the form of short writing and reflection assignments that were built into student notebooks (described later). These assignments asked students to explore the potential relevance of engineering activities to their lives or to the lives of middle school students. Given our focus on developing motivations and capacities to participate in STEM in low-income youth, we drew on this prior work in two areas. One area considers way to develop interests and motivations within our target population. The second area considered how to conceptualize and measure competencies in engineering within an afterschool context.

Motivation and Engagement for Low-income Youth

As Popham[7] states, "student affect—the attitudes, interests, and values that students exhibit and acquire in school—can play a profoundly important role in students' post school lives, possibly an even more significant role than that played by students' cognitive achievements." In recognition of the importance of these "non-cognitive" factors in students' academic success,[8] there has been considerable research into a wide array of interventions designed to encourage student interests and motivation in STEM and other academic disciplines, including the previously discussed utility value, value affirmation,[9] and mindset.[10] In this work, we integrated utility value interventions into the UAV curriculum as prior research indicated that this type of approach can be particularly effective with low-income youth.[11]

Utility Value

The concept of *utility value* refers to the perceived usefulness of a topic or activity with respect to an individual's short- and long-term goals. A utility value intervention typically asks participants to reflect on the utility of a recently completed activity or course for their own lives or the lives of others similar to them. For instance, Hulleman et al. (2010) asked students to write a short essay reflecting on the utility of a course in which they were all participating. They found that the simple essay writing intervention increased students' perceptions of usefulness and interest, especially for students with low expected or actual performance. Gaspard et al. (2015) studied a different intervention where they asked students to reflect on and rate how much they agreed or disagreed with a series of quotes reflecting on the utility of a course or topic, where each quote was attributed to a person from a similar background. The basic premise of these types of interventions is that they force students to construct relevant relationships for themselves, which in turn, help to spark initial interest and motivation. For the most part, these interventions have been used with high school and college students engaged in required coursework. The results have been quite remarkable. This research also observed improvements in students' grades within the course. Beckett et al. (2009) have also studied this intervention with low-income youth in science classes; results suggest it improved both motivation and student learning outcomes and promoted persistence in STEM courses in subsequent years.

Learning/Competency in Afterschool Programs

Competencies have been defined for educational contexts as, "… complex ability constructs that are closely related to performance in real-life situations."[12]

Measuring competency in afterschool programs is a challenge, particularly for topics as large and complex as engineering. Building on recommendations from both Accreditations Board for Engineering and Technology, Inc. (ABET) and the Next Generation Science Standards (NGSS), we decided to focus on understanding and measuring students' ability to use and apply the "engineering design process."

Engineering design is an iterative, decision-making process in which the basic sciences, mathematics, and engineering sciences are applied to optimally convert resources to meet a stated objective. Among the fundamental elements of the engineering design process are the "establishment of objectives and criteria, synthesis, analysis, construction, testing and evaluation."[13] According to the National Research Council's K–12 Framework for Science Literacy,[14] there are both practical and inspirational reasons for including engineering design as an essential element of science education:

> We anticipate that the insights gained and interests provoked from studying and engaging in the practices of science and engineering during their K–12 schooling should help students see how science and engineering are instrumental in addressing major challenges that confront society today, such as generating sufficient energy, preventing and treating diseases, maintaining supplies of clean water and food, and solving the problems of global environmental change.

Building on the notion of competency as "related to performance in real-life situations," we are drawing on classroom observations and analyses of student work. We also developed several design "challenge" questions to better understand the degree to which students in the *Engineering Experiences* program were able to generalize from their UAV engineering design activities to other engineering problems. These design challenge questions asked students to talk through how they would go about designing either a rubber-band powered car or wind-powered sailcar that travels an optimum distance.

Context for this Work

The school selected for this study is a middle school located in Colorado. The school has an emphasis on academic achievement and social learning but also strives to assist students in learning beyond specific content to include communication, human relations, problem solving, critical thinking, and collaboration. The school offers a pre-engineering program, which feeds into the high school engineering track. Based on Colorado Department of Education statistics from 2014, the study school is comprised of 562 students with 52%

qualifying for free or reduced lunch. Within this school site, we implemented two deployments of the UAV curriculum with two different student cohorts (See Table 6.1).

Cohort one was recruited from the *I Have a Dream Foundation* afterschool program. The *I Have a Dream Foundation* is dedicated to helping low-income youth achieve a brighter future through a long-term, comprehensive educational, and cultural enrichment program; they also provide wraparound services for families. The criterion for selection into the program is that the student is on the federal free- or reduced-lunch program or s/he lives in a low-income housing site. Formed in 1990, the local chapter of *I Have a Dream* operates multiple after school learning centers and manages a distributed network of mentors drawn from the local community and businesses. Approximately 85% of *I Have a Dream* graduates continue on to post-secondary education. Based on conversations with the leadership from the *I Have a Dream Foundation* program, we worked directly with one of their sites, which serves predominantly Latinx middle school students with our new UAV/Drone curriculum.

Based on the OST framework, we were very intentional with connecting the experiences to support the instruction going on during the school day at the middle school. This involved conversations with the engineering teacher, members of the science department and the principal. It was during one meeting with the principal requesting permission to fly UAVs in the gymnasium after school, that we learned about a Saturday school program that he runs targeting the other students from low-income families from the same community but not necessarily enrolled in the *I Have a Dream* program. He was intrigued with our curriculum and after learning about our struggles with consistent attendance, suggested that we work with his teachers to offer the UAV lessons on Saturdays. After a short professional development with the teachers, they were excited about the possibilities of integrating the UAV lessons with their Saturday programming.

Our second cohort involved students participating in the Saturday school program. The Saturday school program was open to all students at the middle school and has traditionally been a time for students to come in for extra instruction in mathematics and language arts. The program is run by three 'master teachers' and volunteers. The *Engineering Experiences* UAV/Drone curriculum was integrated into the program in a way in which the language arts and mathematics lessons made use of the context of the student experiences using the UAV/Drone activities. The faculty promoted the new addition by showing an informational video to students during the weekly announcements leading up to the spring 2017 implementation.

Table 6.1: Key features of the cohorts and settings for the two UAV curriculum deployments.

Feature/ Deployment	Cohort One: I Have a Dream Program- MS in Colorado	Cohort Two: Saturday School- MS in Colorado
Recruitment	Students selected UAV experiences from several other after school options including relaxation, sports, and homework help.	Students signed up to attend this program knowing that they would participate in UAV lessons and assistance with language arts and mathematics.
Time	Fall 2016 Afterschool Tuesdays or Thursdays from 5:15–6:00 PM	Spring 2017 Saturdays for 60 minutes between 9:00–11:30 AM
Adult Leaders	STEM coaches from U. Colorado engineering program with IHAD volunteers	Curriculum developers from UCAR with a teacher and volunteers
Curricular features	Ten sessions with a culminating simulated aerial survey of the fictitious town of 'Disasterville.'	Nine sessions with a culminating lesson 'Disasterville' followed by a rescue planning and implementation.
Attendance	~4–12 students each day per week	~40–50 students each week
Incentives	Weekly attendance incentives included content-based giveaways (e.g., cloud viewers, tornado tubes) and a drawing to win a UAV/Drone to take home for students with highest attendance.	Each year student attendance is incentivized with a weekend trip to the mountains.
Data Collection	Flight logs, observations, interviews	Flight logs, observations, interviews
Competing Options	Students had several options for activities (sports, crafts) in addition to *Engineering Experiences*	Students attended the Saturday School for the expressed purpose of help in mathematics and language arts in addition to *Engineering Experiences*
Utility Value Intervention	Students were asked to write a short essay reflecting on the relevance of engineering to their lives, based on Hulleman et al. (2010)	Students were asked to reflect on various quotes about engineering and to rate the relevance of these quotes to their lives, based on Gaspard et al. (2015)

Source: Author.

Curriculum Design

UAVs, especially inexpensive models suitable for educators' budgets, are a very recent innovation. Though UAVs are a very new on the scene, they appear to have quickly captured the interest of many formal and informal educators.[15–19] At present, there are not many detailed curriculum to support use of UAVs at the K–12 level. Some detailed lesson plans are available,[20,21] along with less developed nuggets of ideas for lessons such as "determine the speed of a UAV over a fixed course" or "record video of a school sporting event."[22] The curriculum we employed drew on a few ideas from existing materials, though most of our UAV curriculum was developed explicitly for our project.

The UAV curriculum involves students in a series of activities over a ten-week period leading up to culminating challenge activity: conducting an aerial survey of a remote mountain town that suffered unknown damage in a storm. The curriculum has three main segments that include seven lessons. Two additional lessons were developed after initial testing of the curriculum with the first student cohort. Student "flight log" booklets, somewhat akin to journals, were used throughout the curriculum to fulfill a couple of needs. In concert with the teachers who were running the second cohort program, we developed mathematics, literacy, and technology extensions that linked the teachers' standard curriculum with the UAV activities.

Figure 6.1: Notebook (Left) or "Flight Log" Used By Participants and Prompts (Right) Pasted on The "Flight Log."
Source: Author.

The first of the three segments of the UAV curriculum can be characterized as "UAV pilot training." Students must first learn to fly the UAVs as a prerequisite to performing subsequent science or engineering activities with the aircraft. Students learn basic terminology about aircraft (pitch, roll, yaw) which they use to communicate to each other while flying; learn the operation of the joysticks and buttons on the UAV controller; and practice increasingly challenging flights to learn how to take off, land, and maneuver in-flight.

After the students have learned to be successful flyers, they conduct a pair of scientific investigations to measure UAV performance. Both experiments provide data about the UAVs' capabilities that support planning of later "missions." They also help transition students' mindsets from "we're flying drones" to "we're conducting investigations that include data collection with drones." Students measure the battery lifetime of the UAVs during flight (about eight to ten minutes with no payload), which determines the possible duration of a mission and the potential range of the UAV. In another lesson, students progressively attach small weights (washers) to the UAVs to determine how heavy of a payload the aircraft can carry. This "carry payload" activity introduces the first small taste of engineering design to the UAV lessons; students use their choice of pipe cleaners, rubber bands, etc. to attach the weights to the UAVs. Students discover that symmetry and balance are important when attaching payloads, and that a payload that dangles down can sway in flight, significantly altering the performance and ease of control of the UAV.

The final two lessons present students with design challenges. In the first design challenge lesson, students must use their UAVs to retrieve a payload from the far side of the gymnasium and return it to a target landing zone. The students design "sky hooks" to attach to the UAVs that "grab" the payload, again using their choice of rubber bands, pipe cleaners, paper clips, tape, and the like. It is exceptionally challenging to grab the payload with a simple mechanical hook, such as a bent paper clip hooking a loop on the top of the payload. We place magnets inside the payloads, and provide students with magnets to incorporate into their sky hooks, to make the challenge more reasonable for students to complete successfully. Since students tend to initially build elaborate but not especially functional sky hooks, this lesson provides a good opportunity for students to practice the engineering design cycle by testing and modifying their sky hook designs in an iterative fashion.

The second design challenge requires students to use the UAV's camera to survey a mock disaster area. The UAVs we used include small video cameras that provide a live feed to a smartphone or tablet. We built a pretend town from blocks and toy cars and small figurines, then hid it from student view behind a low "mountain range" (a plastic tarp draped over some chairs). We wrecked some portions of the "town" to represent damage caused by a disaster (tornado, flood, etc.), and provided students with a map and photos of the town in its pre-disaster state. Since students could not see the town (we called it "Disasterville") directly because of the intervening mountain range, they had to fly their camera-bearing UAV over the town to survey the extent and location of damage.

Figure 6.2: Photo of Disasterville Scenario.
Source: Author.

This "Disasterville" scenario incorporates two engineering aspects. The standard "factory" mounting of the video camera on the UAVs has the camera pointing horizontally, such that the camera looks ahead in the direction the UAV is flying. To survey Disasterville, students need a downward-pointing camera. Students were again provided with tape and rubber bands and similar supplies, and proceeded to "hack" the camera mounting to make it point downward. This task presented constraints that were provided to students for consideration before they began their designs: the camera has a short wire that must stay connected to a specific spot on the UAV, the radio antenna on the camera must be kept clear of the UAV's propellers, and the camera must be mounted in a way that doesn't disturb the aircraft's balance too much. The second engineering element of this activity involves mission planning. The video feed if sufficiently choppy and low resolution, and the "hacked" camera mounting disturbs the UAV's maneuverability enough to make it quite challenging to acquire footage of the entire town. Students must plan their survey flight to sweep over the whole town, getting enough footage to thoroughly survey for potential damage. Students were encouraged to record video during flight, and were given time to replay their footage (pausing and using slow motion to scrutinize key segments in detail). This "image analysis" phase can lead to further iterations in the form of a second flight to survey areas that were missed during the initial reconnaissance or areas that needed further attention (e.g., infrastructure or vehicles that were damaged, people who need medical attention).

After using the curriculum with cohort one, we developed two additional activities related to the Disasterville scenario. One activity is similar to the first design challenge described earlier, in which students design a sky hook to retrieve a payload. In the new activity, students must deliver small buckets representing water to a designated landing zone target near Disasterville. Students once again were supplied with pipe cleaners and rubber bands and the like, which they used to design and build a device to hold the bucket during the aid flight. Designs need to balance the requirements that the buckets remain attached during takeoff and the flight to the landing zone, but must allow the bucket to detach when the UAV touches down in the landing target without any further human intervention. As was the case with the sky hooks in the payload retrieval mission, initial student designs for the bucket carriers were generally unsuccessful, requiring students to iteratively improve their designs.

Throughout the curriculum students used "flight log" notebooks for a variety of purposes. In some activities (such as the battery duration test and the test to see how much weight a UAV can carry) students needed to record data, so the flight log was used as a scientific laboratory notebook. During most activities, students were asked to respond in writing to a printed prompt, either at the beginning of the session, the end, or both. These short writing assignments helped focus student attention at the start of activities or helped students reflect on and process their experiences at the end of the session. As shown in Table 6.1, one of the short assignments provided to students through the flight logs was a utility value intervention. These flight logs also served as an integrated, embedded research instruments to help document student thinking. Since student attendance, especially in cohort one, was inconsistent, and because some lessons are essential prerequisites for others, we needed a way to keep track of which activities each individual student had successfully completed. We used a badging system as is common in recent gamification efforts or reminiscent of the merit badge system used in scouts. Activities, and in some cases subsections of activities, were assigned criteria for successful completion. When students completed a specific task, they were awarded a badge-like certification of their success. Instead of awarding physical badges, we used an inked, rubber stamp to mark in their flight logs whenever they completed a task.

During the Saturday school program of cohort two, one group of students did the UAV curriculum during the first hour, then switched to math or literacy in the second hour; a second group did math and literacy first, followed by UAVs. In several instances, we were able to work with the teachers to integrate aspects of the UAV curriculum with their math and literacy lessons. The UAV activity that tests battery lifetime includes an opportunity to do some simple statistics using the measured lifetimes of the batteries, which are not identical. The math teacher wanted students to practice working with spreadsheets, so battery data from a number of trials were recorded in a spreadsheet, allowing the math teacher to use student-collected data while discussing statistical concepts such as mean and range. The math teacher also wanted to incorporate UAV data into story problems involving time and rates. For example, problems like "How far can a UAV fly on a single battery charge?" or "Could a UAV deliver a pizza five miles away within 20 minutes?" require knowledge of the UAV's speed. To support this, we conducted several speed trials with the UAVs in parallel with other lessons one week, timing several flights over a measured distance. Data were once again entered in a spreadsheet to provide students practice doing calculations with technological tools.

Early in the session, students were shown a short video clip (similar to a movie trailer) introducing them to the "Disasterville" activity scenario they would encounter weeks later. They completed a brief writing assignment in reaction to that preview, and decorated the covers of their flight logs. After conducting the "Disasterville" mission, students brainstormed about what they need to know in order to determine that they can deliver goods and then completed a storyboarding extension to map out their plans. Building on the work of Sadler, Coyle, and Schwartz (2000), students use words and/or pictures to fill in frames of a comic book-like storyboard about their aid mission, explicitly addressing specific engineering issues such as "what problem are you trying to solve?" and "what is your solution?" "The storyboard is a series of frames created by students during the project, each frame displaying the latest solution to the challenge."[23] Based on this work, we provided three prompt options and revised the storyboard template to make the connection to the engineering design process explicit. We provided structure for the storyboards by asking students to respond to steps in the engineering design process. One of the prompts provided to the students was a scenario about delivering medical supplies to Disasterville:

> Based on the aerial survey video of Disasterville, doctors have decided that medical supplies need to be delivered to the town. The UAV can carry medical supplies to a safe landing zone near the town. On your storyboard, design a canvas bag that can be carried by the UAV and can hold medical supplies for delivery to Disasterville.

In one completed storyboard collected from participants, students chose to design a case with handles on the side rather than the canvas bag described in the prompt; most students did not include an explicit improvement to the solution as prompted in the storyboard template. After students developed their storyboards, they presented their ideas to their peers. The debrief provided participants with an opportunity for peers to ask questions and gain clarification about the details of each of the rescue plans and to ask and answer questions. The storyboards were useful to student participants as a tool to connect information gathered by the aerial survey to what it would take to provide real-world assistance to the town. It also provided the teacher the opportunity to challenge students to think about improving their design based on actual trials modeled with the UAV.

Table 6.2: Planning a Rescue Storyboard Template.
Use to draw and/or write the steps of your rescue story in the six boxes below. Use the guide questions to focus your thinking.

Ask: What is the problem?	*Explore: What are some solution ideas?*	*Design: Choose the best solution plan.*
Draw a picture from the UAV video that describes the problem.	Draw a picture or use words/phrases that describe some engineering solutions.	Show how one solution would be designed prior to take off.
Create: Follow your plan.	*Create: Follow your plan.*	*Improve: How did the solution work?*
Show how your solution would work on the way to the town.	Show how your solution would work during the rescue at the town.	Using your imagination, show the result of the rescue. How could you improve?

Source: Author.

Research Design

Our research examined the degree to which the UAV curriculum influenced students' interests and motivation in the *Engineering Experiences* program, their beliefs and attitudes towards UAVs and engineering, and their understanding of the engineering design process.

Participants

Our study involved samples of students drawn from the two different cohorts described earlier. Cohort one consisted of 32 participants (seven female) who were part of the *I Have a Dream program*. Informed consent was obtained for 24 of these participants through outreach during a parent meeting. Cohort two had 55 participants (20 female) who were part of the Saturday school session. The teachers responsible for organizing the Saturday school session helped collect 38 informed consents.

Procedures and Measures

We collected both qualitative and quantitative research data using prompts embedded in students' flight logs, as well as semi-structured interviews and classroom observations. Participants were asked to fill in different prompts every day at the beginning of the session. Table 6.3 shows example prompts and how they are aligned to each of our research constructs.

Table 6.3: Research Constructs, their definitions, and example prompts used in the flight logs.

Construct Name	Definition	Example Prompts
Study Variables		
Initial Interest	Excitement or motivation at the beginning of performing an activity	• I like to learn how things work • I am excited to learn more about engineering • Engineering is boring • I like building cool stuff
Performance Expectation (pre)	Expectation level at the beginning of performing an activity	• I think I will do well in *Engineering Experiences* • I think I will learn something new in *Engineering Experiences* • I do not think I will be learning anything new from *Engineering Experiences* • How do you feel you will perform while taking part in *Engineering Experiences*?
Utility Value	Relevance of the activity to one's life	• Engineering is useful in everyday life • I don't think learning engineering will be useful to me • I think what I will learn in this engineering activity is relevant to my life

(Continued)

Table 6.3: (*Continued*)

Construct Name	Definition	Example Prompts
Situational Interest	Short-term or a momentary interest in a specific task	• Participating in the engineering activity was fun • To be honest I thought the engineering activity was boring • It was a waste of time to learn about the project
Performance Expectation (post)	Expectation level at the end of performing an activity	• I think I am doing well in *Engineering Experiences* • I have learned something new in *Engineering Experiences* • I have not learned anything new in *Engineering Experiences* • How do you feel you are performing in *Engineering Experiences*?
Maintained Situational Interest	Long-term or enduring personal interest to continue pursuing an activity	• I will use what I learned from this engineering activity in the future • I am interested in learning more about engineering
Formative Feedback		
Rose/Bud/Thorn	Ways to improve specific tasks in the curriculum and feedback about the program as a whole	**Rose:** What has been the best thing about *Engineering Experiences* so far? **Bud:** What are you looking forward to about *Engineering Experiences* in the coming weeks? **Thorn:** What has been the worst thing about *Engineering Experiences* so far?
Rant/Rave		Rant about something bothering you about *Engineering Experiences* and/or rave about something exciting you about *Engineering Experiences*.

Source: Authors.

Each student had a personal "flight-log" book or notebook where the prompts were pasted. Participants were asked to rate the degree to which they agreed with the prompts on a five-item scale. Every week a member of the research team would paste in the appropriate prompts in the "flight log" books. At the end of the day, a research team member would transcribe all data, anonymize data, and store it in a secure location. Once participants had completed the initial interest and pre-activity performance expectation prompts, they were provided the opportunity to respond to the utility value task. This task took place at the middle of the semester when participants had some understanding of the curriculum. For cohort one, the participants had to evaluate quotations on engineering. Participants were provided with a list of six quotations about the relevance of engineering to the lives of six people who were interviewed prior to the study. Participants had to mention the degree to which they agree with the quotations and if they had similar thoughts or opinions about engineering. They also had to rate the importance of each of the quotations to their lives. Cohort two was asked to write a short essay (three to five sentences) about the potential relevance of engineering to their lives and to the lives of middle-school students in general. In the next sessions after the utility value intervention, participants completed the situational interest, post-activity performance expectation, and maintained situational interest prompts. Additionally, several prompts asked students questions on drone control and maneuvers like pitch, yaw, and roll, while other prompts asked students to provide formative information to help improve the UAV curriculum and the *Engineering Experiences* program.

We conducted semi-structured interviews at the beginning and end of the program. Students were asked about their views on engineering and UAVs, and we posed an engineering design challenge question and asked them to describe how they would go about addressing it. All interviews were transcribed and coded by two members of the research team. Student responses to the engineering design challenge were coded to identify the design moves proposed by the student, the evaluative criteria or constraints that the student used to assess the quality of their proposed solution, and the reasoning students provided, if any, to justify their design choices. Members of the research team also observed numerous sessions with both cohorts. Our observation protocol focused on assessing student engagement with the activities and the degree to which they were enacting engineering design processes.

Results

We report on our main findings in three areas: changes in students' interest and motivation, changes in their perspectives on UAV and engineering, and their understanding and enactment of the engineering design process.

Changes in Students' Interest and Motivation

Utility value theory holds that students' maintained situational interest should increase as a result of the intervention, at least for some students. There were no significant changes in students' quantitative self-reports of interest, in either cohort, comparing their initial interest with their reported maintained situational interests. This finding stands in contrast to the positive changes reported elsewhere. This result may stem largely from a ceiling effect: students' incoming initial interests in UAVs and *Engineering Experiences* was very high (17 out of 20) based on the cohort response averages, providing little scope for these measures to increase. Qualitative data from the interviews support this interpretation as the majority of students reported that their primary reason for signing up for *Engineering Experiences* was their excitement and interest in drones. This enthusiasm was reiterated in the post interviews where student participants, indicated here using P#, typically offered comments such as:

> It is a great opportunity to fly drones and play with electronic stuff. And right now I feel like it's pretty cool because you get to put cameras on the drones, and you get to fly over something that you can't see. [P2]

> In the beginning it was very hard to fly the drones. And I'm – and I had a little experience but not that much experience. But now I feel like I'm like a total pro. I feel like it's very educational. And I feel like it really helps with engineering stuff. And it's really fun to learn how the – how to fly the drones. [P4]

Perceptions about Drones

While there was no quantitative change in students' self-reported interests levels, qualitative data suggest that the utility value intervention may have influenced youths' perceptions of drones. During post-interviews and in responses to reflection prompts, youth commented on how their perceptions of drones had changed, from regarding them mainly as toys to starting to realize that UAVs can also serve as tools for science and engineering. For instance, one participant stated,

I think it's a nice experience learning how to use drones or like any kind of technology … I care about it cause I think it's interesting how technology works, how drones work, and how drones can become a better thing, and how they can help people, cause I always thought of them as a toy until this. [P5]

Similar to what was stated by P2, the camera activity helped several youth to realize the utility of drones for remote sensing, and thus provided an entry point for encouraging participants to think about drones as scientific instruments rather than toys.

Engineering Perspectives

During our pre-interviews, about a third of the youth (eight out of 21) expressed worries about their ability to participate in engineering activities. By the post interviews, there was a strong positive affective change. All 21 respondents reported positive attitudes about engineering and their ability to participate. As one of the participants [P3] said, "I feel like engineering was going to be hard for me. I never know, like, what do I do? But then later when I started learning, it was much easier." While another participant [P12] said, "I used to think it was really boring but now I think that it's really fun."

Engineering Design Processes

A key goal of the curriculum is to develop students' competencies and appreciation of the engineering design process; in other words, the ability to establish design objectives and criteria, analyze the strengths and weaknesses of designs, construct new designs, test designs, and evaluate their efficacy with respect to established criteria. Implicit in this process is the vital role of iteration: continually revising and improving a design based on performance evidence. While we observed students routinely engaging in these individual practices during the afterschool sessions, we also noticed that some youth were resistant to iteration and did not want to modify their initial designs. The very short duration of the afterschool sessions (45-minutes) might have encouraged this behavior, as youth really did not have sufficient time to engage in thoughtful redesign and testing cycles. We also investigated the degree to which participating in *Engineering Experiences* improved their ability to apply the engineering design process to other design challenges. However, our analysis of the students' responses to the pre and post engineering design challenges revealed no significant differences. It appears that students were not able to generalize from their UAV experiences to other engineering design problems, an idea that could be related to the relative short time frame of the experience.

Conclusions

To date, *Engineering Experiences* has provided middle school students from both low income and diverse populations with opportunities to fly, design, and use UAVs in a variety of problem-solving tasks, building towards a scientific and humanitarian mission (Disasterville). Our research highlights several lessons to build on, and several areas where the program could be strengthened in future iterations.

Building on Youth's Interests

As described earlier, the motivation for creating a program around UAVs was based on a survey of youth in the *I Have a Dream* program. They overwhelmingly expressed a strong desire to work with drones over other topics such as solar energy or wind power. It is important to note that these were, for the most part, different youth than those who participated in cohort one. Yet, their interests results generalized to other youth from the *I Have a Dream* program and to the entirely new cohort two population. Picking a powerful and compelling topic is essential for the success of an afterschool program that is based on voluntary participation, particularly since prior research suggests that interest in the topic is the primary driver of youth attendance.[24]

Unobtrusive Data Collection

Integrating prompts into students' flight log notebooks was an extremely effective mechanism for collecting research data *while supporting student reflections* on their experiences at the same time. In essence, the research design, with the exception of the pre and post interviews, was deeply integrated into the curriculum and inseparable from it. Our observations confirmed that students regarded these prompts as integral to the *Engineering Experiences* curriculum: after the first few sessions they would come into the classroom, retrieve their notebooks, and start responding to the initial prompts with little guidance from the coaches.

This acceptance and participation stands in stark contrast with our prior experiences. For instance, in a previous afterschool research study (NanoExperiences), students had little interest in completing surveys about their experiences. External evaluators from the NanoExperiences project[25] describe problems related to the feasibility of data collection in low-stakes programs noting that OST program assessment is less typical than other types of program evaluation, perhaps lending itself to lower levels of investment by participants. These attributes contribute to the need to do face-to-face data collection and paper-pencil surveys.

In an effort to address this issue, the researchers worked carefully with the developers to create reflective prompts that could be embedded into the curriculum itself. We did this through including one writing prompt or up to five multiple choice questions in student's flight logs at the beginning of each session. By doing so, the students did not perceive these questions as formal surveys but rather a time to reflect on their experiences as part of the instruction. While there were still issues with all students completing each of the prompts, the perception of having too many surveys was never an issue.

Facilitating Deeper Connections to Engineering Design

We are making several changes to the curriculum to deepen students' engagement in the engineering design process. First, we are foregrounding the overarching design challenge—the Disasterville scenario—from the very beginning so that students understand how each activity helps them to plan for and design for that mission. Second, we are revising both our notebook prompts and the instructional tools provided to the coaches to encourage students to engage in more reflection and discussion of the engineering practices and processes they are engaged in. Finally, we are changing the session structure to be significantly longer, replacing the 45-minute sessions with weekly 1.5 hour sessions and work with both the *I Have a Dream* staff and the teachers at the host school to ensure more consistent participation. Not surprisingly, the setting and environment in which the *Engineering Experiences* activities were conducted had a powerful impact on outcomes. The situation for cohort one was quite challenging. Sessions were held at the very end of the students' days (5:15–6:00 PM) when many students were visibly fatigued and these sessions were too short to enable students to iteratively refine their design. Work with a third cohort (fall semester 2017) will allow us to study the effectiveness of these changes on youths' competencies, skills, motivation and interests in engineering.

Discussion questions

- How do out-of-school STEM programs promote the use of standards in informal settings?

- Identify ways to promote literacy in STEM courses that reinforce learning through modeling and simulations.
- Why is picking a powerful and compelling topic a source of motivation? What topic of study motivates you to engage in a STEM community of practice?

Notes

1. I struggled here with what term to use for the target group. SEEK participants are primarily students of color, and the program is housed in economically high-need school communities. However, "high-need" and "students of color" are not synonymous, though they are often used interchangeably.
2. Institute of Education Sciences and National Center for Education Statistics, 2016.
3. Beckett et al., 2009.
4. Hulleman & Harackiewicz, 2009.
5. Hulleman, Godes, Hendricks & Harackiewicz, 2010.
6. Gaspard et al., 2015.
7. Popham, 2009.
8. Yeager et al., 2014.
9. Sherman et al., 2013.
10. Paunesku et al., 2015.
11. Harackiewicz et al., 2016.
12. Hartig, Klieme, & Leutner (Eds.), 2008.
13. Accreditation Board for Engineering and Technology (ABET), n.d.
14. National Research Council, 2012, p.9.
15. Armentrout, Carnahan, Crowley, & Zieger, 2016.
16. Crook, 2017.
17. McGillivary et al., 2016.
18. Mooney, 2016.
19. Olds, 2016a; 2016b; 2016c.
20. Carnahan, Zieger, Fernandez, Sheehy, & Crowley, 2016.
21. Mooney, Olds, Dahlman, & Lewis, 2017.
22. Carnahan et al., 2016; Mooney et al., 2017.
23. Sadler, Coyle, & Schwartz, 2000.
24. Akiva & Horner, 2016.
25. Additional details about the NanoExperiences project, program evaluation and assessment instruments are available at http://stelar.edc.org/projects/12681/profile/nanoexperiences-pathways-workforce-success

References

Accreditation Board for Engineering & Technology (n.d.). Definition of design. Retrieved from http://myweb.wit.edu/jamesons/Courses/Mech%20Design/lec3.pdf

Akiva, T., & Horner, C. G. (2016). Adolescent motivation to attend youth programs: A mixed-methods investigation. *Applied Developmental Science, 20*(4), 278–293.

Armentrout, J., Carnahan, C., Crowley, K., & Zieger, L. (2016). Panel at the ISTE (International Society for Technology in Education) 2016 Conference, Denver, Colorado; 29 June 2016.

Beckett, M., et al. (2009). *Structuring out-of-school time to improve academic achievement.* Washington, DC: US Department of Education, NCEE 2009–012. Retrieved from https://ies.ed.gov/ncee/wwc/Docs/PracticeGuide/ost_pg_072109.pdf

Carnahan, C., Zieger, L., Fernandez, N., Sheehy, L., & Crowley, K. (2016). *Drones in education: Let your students' imaginations soar.* Arlington, VA: International Society for Technology in Education.

Crook, A. (2017). A STEM approach to integrate drones as a teaching and technology tool. National Science Teachers Association 2017 National Conference, 31 March 2017.

Dahlman, L., Olds, S., & Mooney, M. (2016). Got a drone? Try this ... Learning activities and science fair project suggestions for you and your recreational drone. Retrieved from http://wiki.esipfed.org/index.php/File:Got_a_drone%3F.pdf

Gaspard, H., et al. (2015). Fostering adolescents' value beliefs for mathematics with a relevance intervention in the classroom. *Developmental Psychology, 51*(9), 1226–1240.

Harackiewicz, J. M., Canning, E. A., Tibbetts, Y., Priniski, S. J., & Hyde, J. S. (2016). Closing achievement gaps with a utility-value intervention: Disentangling race and social class. *Journal of Personality and Social Psychology, 111*(5), 745–765.

Hartig, J., Klieme, E. & Leutner, D. (Eds.) (2008). Assessment of Competencies in Educational Contexts: State of the Art and Future Prospects. Gottingen: Hogrefe & Huber.

Hulleman, C. S., & Harackiewicz, J. M. (2009). Promoting interest and performance in high school science classes. *Science, 326*(5958), 1410–1412.

Hulleman, C. S., Godes, O., Hendricks, B. L., & Harackiewicz, J. M. (2010). Enhancing interest and performance with a utility value intervention. *Journal of Educational Psychology 102*(4), 880–895.

Institute of Education Sciences and National Center for Education Statistics. (2016). Table 204.10: Number and percentage of public school students eligible for free or reduced-price lunch, by state: Selected years, 2000–01 through 2013–14. Retrieved from https://nces.ed.gov/programs/digest/d15/tables/dt15_204.10.asp?current=yes

Kefauver, S. C., Sanchez Bragado, R., El-Haddad, G., & Araus, J. L. (2016). Open source software and low cost sensors for teaching UAV science. Proceedings of the AGU Fall Meeting 2016; session ED33F-0920; 14 December 2016.

McGillivary, P. A., Lukaczyk, T., Brendan, B., Tomita, M., Ralston, T., & Purdy, G. (2016). Incorporating Unmanned Aircraft Systems (UAS) into High School Curricula in Hawaii. Proceedings of the AGU (American Geophysical Union) Fall Meeting 2016; Session ED33F-0924; 14 December 2016.

Mooney, M. (2016). Using Recreational UAVs (Drones) for STEM Activities and Science Fair Projects. NSTA 2016 Minneapolis Area Conference; 28 October 2016

Mooney, M., Olds, S., Dahlman, L., & Lewis, P. (2017). ESIP Education Drones in STEM Initiative. Proceedings of the 97th American Meteorological Society Annual Meeting, 26th Symposium on Education, Themed Joint Session 2: Remote Sensing Education Experiences.

Olds, S. (2016a). Using recreational UAVs (Drones) for STEM activities and science fair projects. NSTA 2016 STEM Forum & Expo; 28 July 2016.

Olds, S. (2016b). Using recreational UAVs (Drones) for STEM activities and science fair projects. NSTA 2016 Portland Area Conference; 11 November 2016.

Olds, S. (2016c). Using recreational UAVs (Drones) for STEM activities and science fair projects. NSTA 2016 Columbus Area Conference; 1 December 2016.

Paunesku, D., Walton, G. M., Romero, C., Smith, E. N., Yeager, D. S., & Dweck, C. S. (2015). Mind-set interventions are a scalable treatment for academic underachievement. *Psychological Science, 26*(6), 784–793.

Popham, W. J. (2009). All About Assessment/Assessing Student Affect. *Educational Leadership, 66*(8), 85–86. Retrieved from http://www.ascd.org/publications/educational-leadership/may09/vol66/num08/Assessing-Student-Affect.aspx

Sadler, P. M., Coyle, H. P., & Schwartz, M. (2000). Engineering Competitions in the Middle School Classroom: Key Elements in Developing Effective Design Challenges. *Journal of the Learning Sciences, 9*(3), 299–327.

Sherman, D. K., et al. (2013). Deflecting the trajectory and changing the narrative: How self-affirmation affects academic performance and motivation under identity threat. *Journal of Personality and Social Psychology, 104*(4), 591–618.

Yeager, D. S. et al. (2014). Boring but important: A self-transcendent purpose for learning fosters academic self-regulation. *Journal of Personality and Social Psychology, 107*(4), 559–580.

SECTION THREE

Transformation

Transgressive Practices along the Journey

Section Three is on Transformation. Barnes-Johnson's chapter, Engaged Interdisciplinary Literacy: Research & Practices of Secondary STREAM, analyzes the use of lesson study as research-into-practice methodology. Insights were gained by using literacy techniques in a secondary physical science course designed to meet the needs of students from underrepresented groups over several years. Course goals of activating creativity and developing STEM identity in low performing students were met while working to sustain STEM interests of traditionally successful students. The chapter examines ways of unpacking course designs to meet learners' needs, and invites collaborators—other teachers—into the conversation about science education reform in the classroom.

Suess, Chae, and Lewis describe how health education programs can be used as a motivation to enter the STEM pipeline. Their work with teachers and high school students to develop health literacy curricula is an example of STEM enrichment that expands and supports physical and life science education. The health literacy curricula educate high school students about fundamental principles of medicine and specific diseases that affect their community. The researchers also strive to improve healthcare and health literacy by improving communication and cultural competence for both high school and medical school/undergraduate student populations and increasing participation by members of non-dominant communities in health-based careers and post-secondary education. Findings address the merits of the program and lessons learned regarding implementation strategies that could support replication efforts.

Johnson, Schendel, McClellan Ribble and Liu foreground student voices in research undertaken at an urban commuter university. This chapter is an outcome of research conducted under an institutional capacity building grant that utilized an Equity Scorecard[1] report as its framework. The larger study's overarching research question examined what various program stakeholders viewed as key elements of a scholarship program design. Data collection was conducted with university faculty, student support services providers, STEM students, community college faculty, community-based organizations, established programs funded by the same grant, school/district partners, students at partner secondary schools, and School of Education administrators. All stakeholder groups provided detailed and varied recommendations for the new program design. This chapter specifically focuses on the voices of the university students in STEM majors, some of whom are pursuing teacher licensure.

Note

1. The Center for Urban Education at the University of Southern California provides resources for equity process development and data analysis. See https://cue.usc.edu/tools/the-equity-scorecard/ for additional details.

CHAPTER SEVEN

Engaged Interdisciplinary Literacy

Research and Practices of Secondary STREAM

JOY BARNES-JOHNSON[1]

Abstract

Embracing lesson study as a viable research-into-practice methodology, this chapter presents four stages of lesson study designed to meet the needs of curriculum design in a high performing district. Each stage of the process presented the author with opportunities to reconcile her desire to be both progressive and transgressive in her thinking about teaching and learning. This paper reports theoretical and practical insights gained by using literacy techniques in a newly designed secondary physical science course over several years. Implemented to meet the needs of students from underrepresented groups (language diverse students, students with specialized learning needs including assistive technology, and generally disengaged students), an intended function of the class was to activate creativity and STEM identity in low performing students while sustaining STEM interests of traditionally successful students. This paper reflects the results of the experiment

---

1 Joy Barnes-Johnson, Science Educator
 Princeton (NJ) Public Schools
 joybarnesjohnson@princetonk12.org

undertaken to understand course design to meet these critical needs. The paper is organized by first sharing insights at the Stage 1 level: STEM curriculum reform. The original 2012 –2013 course plan (Stage 2) including assessment and differentiation activities will then be shared. Based on student work/artifacts, rethinking required to adapt to the changing profile of the school and science education in general (Stage 3) will be shared. The implications section of the paper is an open invitation to expand the experiment and invite collaborators—other teachers— into conversation about science education reform in the classroom (Stage 4). Lesson plan outlines are shared as a response to the need for researchers to make their findings and recommendations[1] completely accessible to the practitioners who seek to use them.

Editorial Reflections

Olympic-level gymnastics is one of my, Joy's, favorite spectator sports. I don't remember now which year of competition it was that inspired me to think about using a similar system for grade differentiation. Patterning grading after the example provided in competitive sports—a truly authentic model of performance assessment—makes sense. Imagine this illustration: a gymnast is the last up for team competition and her team is tenths of a point behind the leading team. In that moment, she and her coach decide which "test" to take. Should she demonstrate her prowess by fully realizing excellence showcasing a skill or routine that has a lower point value but higher degree of accuracy? Or, does she choose to do a harder skill with a higher point value and lower degree of accurate completion? Inspired by the gymnast who chose the latter path, I crafted a policy for classroom evaluation. I made tests that had a maximum value of 100 points and alternatives whose maximum values were lower, creating two or three versions. The results were less than inspired. Students who accepted the challenge and performed at low levels, begged to redo their tests. Students who took the more "accessible" tests successfully, stopped challenging themselves with harder pursuits. Those who took the more "accessible" versions unsuccessfully, expected additional options. I abandoned the practice after only a few trials, realizing that there needed to be a philosophical change in approaches to assessment in my class but also more holistic alternatives to testing in general, especially as practiced in this environment.

Differentiation battles are real, even among colleagues who seek "fairness" over equity. Discussions about what is fair drive practice and blind the well-intentioned. Equitable schooling however, does not (or should not) mean easy. Equitable assessments are better described as "appropriate." Differentiating instruction so that students have opportunities to demonstrate their understanding in a variety of meaningful ways is important. In order to make this happen, new literacies and novel media offer hope to teachers. In terms of the reality of our "multi-literate world," we can expect students to use traditional tools of literacy in most STEM classrooms. Reading books and journals, keeping notes and writing reports, listening to lectures, observing new phenomena and speaking at symposia are the traditional pathways of receiving and distributing information about STEM. In 2017, students are in fact following/honoring those same traditions, except now, they are going beyond paper to do it. The media platforms that students use to engage in STEM discourses are multimodal and embrace cultural markers like never before. Nerd culture (which includes a fantastic world of animation and gaming), Hip-Hop culture, and youth culture in general have created opportunities for broader participation, better evaluation and improved relationships in my class. A fuller examination of these inclusive practices embedded in my class is relayed in the upcoming chapter.

I, Janelle, admit that I felt a tinge of envy reading Joy's upcoming chapter as she describes policies and structures at the institutional level that facilitate transgressive teaching. As a field and as a profession, for many of us it feels like we have gone backward in many ways and that structural supports for teachers and students in public schools are being intentionally and systematically dismantled. One of the enduring structures in teacher preparation and certification in most U.S. states is the track teachers must choose between elementary and secondary education. Because of my time teaching outside the U.S., I have been lucky in my K-12 teaching experience to teach math and science at the elementary, middle, and high school levels. Each level was humbling in its own way, and I always thought about how elementary and secondary teachers could benefit from each other. If my resources were limitless, I would pay for substitutes so teachers could visit other schools and classrooms. Since that seems to be impossible, unfortunately, I do my best to cultivate the kinds of learning that would come out of an experience like that into teacher preparation work. There is much research literature problematizing the STEM content knowledge of elementary teachers, often representing a deficit view. This same deficit view is reflected by university-level providers of professional development for teachers. While it is true many of the teachers do not have specialist-level knowledge of each field of STEM, many elementary teachers are incredible pedagogues.

In Colorado, secondary school teachers major in their major content area as they work towards teacher licensure. We know from the research that we most often teach as we were taught. For secondary teachers, their primary teaching influences are their content area professors, many of whom have little background in pedagogy. This can be very problematic, especially in terms of equity and inclusion—most professors typically lecture, they rarely differentiate their instruction or use cooperative learning, and heavily weigh summative assessments that have a weeding out effect on students who have experienced educational opportunity gaps. Another struggle we face in our work in Colorado is embedded in the siloed nature of secondary teacher preparation. Although the state does offer a STEM teacher certification, that is limited to Career and Technical Education

(CTE) track courses, not for college prep. Students and their parents who have their sights set on college steer away from integrated STEM classes, though most of the STEM classes I have observed would far better prepare students for real life work than much of the textbook-based coursework, including those found in AP classes.

While many of us feel that the shift toward more integrated STEM and problem-based learning is coming, it is not here yet. There is often a long lag time between invention and execution of school-related plans. It takes years for innovative ideas to find their way into state standards, and then into actual implementation in schools. Higher education seems to lag even further behind, clinging onto its hallowed silos. The space for transformation we have found in our work with pre- and in-service teachers is in unit planning and lesson planning. Parallel to some of the curricular and assessment experiences outlined in the upcoming chapter, students choose community-relevant problems and shape them into summative performance assessments. We help them backward design from there, digging into the standards and shaping essential questions. We push the students to become transgressive teachers who differentiate their lessons and plan ongoing assessment that engages their students, rather than weeds them out.

The goal of developing interdisciplinary curricula that provide seemingly boundless opportunities to learn STEM is beyond ambitious and not without its problems. As with any technology or tool, a curriculum is only as visionary as its designer(s) or as applicable to new purposes as the consumer(s) who uses it. As part of this reflection, we—Joy and Janelle—discovered glitches in our own experiences as editors that expose these limits. The risk of changing the meaning of things simply because there are limits to the tools we [all] use to communicate and collaborate, including time and personal history/exposure. We have been and remain resistant to the idea of a fallacy of linearity[2] with regard to STEM curriculum design. There is no need to conceptualize equitable science teaching as a brick-by-brick experience with only one way to address these challenges. This chapter does not take for granted these limits or expect to serve as the only "good" example of interdisciplinary design.

Introduction

School-based curriculum design to meet the needs of heterogeneous student populations is challenging, especially in a high performing district. Distinct from low performing districts, resource allocation at high performing schools is typically related to how teachers enact reforms.[3] In a study of five high performing schools, researchers recognized that schools in collaboration with state and local efforts were representative of sound reforms. The reward for this cooperation is greater outcomes for students and teachers—key stakeholders and torchbearers for excellence in the school. Five key principles at play were: (1) reduction of specialized programs, (2) more flexible student groups, (3) structural support for relationship building, (4) varied blocks of instructional time, (5) more common planning time. These principles are common in Princeton Public Schools (NJ). Princeton High School is consistently ranked highly among schools in New Jersey by various indices. Consistent with census data,[4] the growing high school has almost 1,600 students and is easily classified as highly diverse.

The mission and educational philosophy of the Princeton Public Schools[5] inspires the entire school community. With the vision "to prepare all students to lead lives of joy and purpose as knowledgeable, creative, and compassionate citizens of a global society" teachers and school stakeholders are tasked with the challenge of using the best information to support the collective strength and efficacy of the community—partners include local universities, libraries, industries and citizens. District leadership and stakeholders openly acknowledge that realizing this vision for all students requires partnerships, innovation and care. Every year teachers are greeted with a convocation that leaves us charged to do this work feeling mostly supported even though it is challenging. We are given multiple mechanisms for professional growth including planned whole-district/whole-school professional development days and are expected to work with peers within and across departments. This expectation comes with built-in planning time (daily and weekly), an affordance that is part of the district's resource allocation plans. Thus there exists infrastructural support for equity-minded thinking.

At the classroom level, it is not uncommon for a full range of students unique in their gender/gender identity, cognitive ability, geography, family socioeconomic class, family composition, race, primary operational language, belief systems and, able-bodiedness to be seated in the same classroom at the same time; even when class composition seems homogeneous, it is not. For example, it is not uncommon to have an honors class with several students with identified disabilities, recommended 504 plans and legally binding individualized educational plans (IEPs).[6] It can also be expected that a special education class could be populated with students who have very high interest, motivation and capacity to do science well at the same rate as those who have little interest, motivation or early exposure to science. These diversity indices contribute to the complexities associated with implementing change in support of equity in the STEM classroom.

After completing my dissertation in 2011, I found myself well supported to implement findings from my own and others' research about equitable practices in science classrooms. When coupled with the 2012 emergence of the Next Generation Science Standards into the public discourse[7] about STEM education reform, my districts' core values of "innovation, creativity and risk-taking" felt like an open invitation to be progressive and transgressive in my approach to teaching. The newly operational professional learning community (PLC) structure for professional development created the space for me to collaborate with peers in my department who teach chemistry to examine our curriculum and imagine what could come next for students. Many students in our school take chemistry as 10th graders and go on to take traditional routes to accelerated science; partnerships with local stakeholders make this possible for motivated students. However, I was becoming aware of a population of students who rarely participated in advanced formal or informal science activities: Black and Brown students. The lack of third-year alternatives to traditional course pathways in physics or the life sciences was equally problematic for me; having few options for students to explore applications of chemistry that were more interdisciplinary within the fields of science and not necessarily "advanced"[8] was an issue. Missing from our course catalogue was a general STEM education alternative that would be appropriate for any student to take classes that would contribute to their joy, their knowledge, their creativity, their passions and their participation as active global citizens in an expanding world.

Stages of Lesson Study

Adapted from a Japanese model of teacher professional development, lesson studies are being used in a variety of ways in the United States. A primary way is in the improvement of teaching for in-service teachers. In the case of planning for diverse classrooms, practicing teachers and administrators often work together to determine what modifications need to be adopted to achieve success for students. This "working together" can be a contrivance of mandates by school, district or state policy shifts or authentically based on shared vision and cooperation. In the case of the former, it might be expected that there is little buy-in or sustainable culture for commitment to reform. In the latter case, collegiality, trust and mutual respect create opportunities not possible in the former. According to Lewis, Perry and Murata (2006) a lesson study cycle is undertaken in stages based on an identified need. Traditionally, there are four stages of a lesson study: (1) studying curriculum for the purpose of formulating goals; (2) planning instruction simultaneously with assessment tools;[9] (3) conducting research using stage two elements as treatments for topics of concern; and (4) reflecting with others about the experiment and lesson revision based on outcomes. This model was well aligned to the professional learning context in the growing and rapidly changing district where NGSS were being adopted and philosophies on STEM education were evolving.

Stage 1: Studying Curriculum and Formulating Goals

The expectation that teachers would work together to establish PLCs was normalized among district leaders and teachers in the science department within my school community. Given autonomy and expected to be creative, teachers in my district have generally been esteemed as professionals. As early adopters[10] of the Next Generation Science Standards (NGSS) teachers readily share resources demonstrating a commitment to differentiation and equity in curriculum design. However, as we have negotiated the standards more, we started to realize the specialized demands of the NGSS to engage students in novel ways. We recognized problems in our own implementation; even as we sifted through the research-based resources available to inform our practice, we were rarely able to reproduce positive results or generate the same level of optimistic outlooks we were finding reported for every performance band we served. The classroom contexts and student group vignettes referenced in the Appendix D Case Studies[11] were familiar to us. Many of the strategies were part of our practice. Even still, we found differing levels of engagement and participation by students across clear lines of home language, race, class, able-bodiedness and ability.

A focus question for professional growth emerged from this need. As a physical science teacher I wanted to know how to develop student success in chemistry-related courses. The current metric for success, ability to qualify for advanced placement courses in a second year of instruction was the beginning. Over time, I realized that the NGSS provided the mechanism to decipher success: application of cross-cutting concepts and observed execution of appropriate practices from the fields of science and engineering to meet the needs of humans, albeit for social, emotional or physical growth and wellbeing, science, technology, engineering and mathematics education is in service to humanity.

> There are three spheres of activity for scientists and engineers: investigating, evaluating and developing explanations and solutions ... The goal of science is to develop a set of coherent and mutually consistent theoretical descriptions of the world that can provide explanations over a wide range of phenomena. For engineering, success is measured by the extent to which a human need has been addressed. (NRC,[12] 2012, pp. 45+)

Narrowing the content focus to the physical sciences allowed me to design a course that could effectively be used as a second year course but perhaps be robust enough to be integrated into the curriculum as a fully compliant capstone[13] course.

Stage 2: Planning Instruction and Assessment.

There are primarily two forces that drive science and engineering: curiosity and practicality. These two forces were used to guide teaching and learning about how science and mathematics shape our society from historical and contemporary perspectives. Defining important relationships between science and math in an age of information and terrorism was an important goal for me as a teacher. Like generations before me, pivotal moments in history around war and violence have come to define how we interact as humans[14] in profoundly dangerous ways. My intention with this class was to purposely attend to the sociological implications of racism, xenophobia, meritocracy and other systems of bias at work in STEM. This was risky, transgressive, and nuanced to some but highly intentional. The Science Technology Engineering and Mathematics (STEM) in a 21st Century Society (STEM-21) class was designed to be a physical science elective lab course for students who have completed at least two years of lab science. Science and mathematics content were presented as primary content foci and anchors for each of five units (four direct content instructional units and one independent research unit), somewhat like what could be imagined as bookends for a shelf of important knowledge. To demonstrate understanding of these focal ideas, learners would have to create models to explain the central scientific tenets and related sociological implications of those ideas using quantified data. Students could use engineered physical or simulated technological tools to create these models. Knowing

Table 7.1: STEM-21 Unit Curriculum Design.

Direct instruction	Unit I: Design for Disaster	Unit II: Forces, Motion, Energy & Structure	Unit III: Human Impact	Unit IV: Alternative Energy
Physical science concepts	Matter Materials Geography Engineering design process	Newtonian physics Energy types Engineering design process	Mapping & geography Gas laws Mineralogy Nuclear energy Engineering design process	Organic chemistry basics Forms of energy (renewable and non-renewable) Mapping Engineering design process
Math concepts	Measurement & unit conversion Mapping (Area, Spatial relationships) Number systems Scale & proportion	Algebraic reasoning	Data collection and analysis (primary and secondary)	Public data analysis Personal consumer data analysis

Source: Author.

that in many cases students' ideas would perhaps go beyond my own skill set, as the teacher, I exchanged *my* pedagogy for *our heutagogical* practice; moving from personal theories about instruction into a more holistic and corporate belief about teaching and learning, key shifts in my teaching increased student participation. Heutagogy as "self-determined learning" provides students and teachers in STEM classrooms with key transgressive tools: purpose, self-efficacy, engagement, voice, visibility and accountability.[15]

A major content theme for this course is the material world and design choices within it. Students were expected to use various technological applications as the vehicle for guided inquiry, independent research and communication of understanding. Imagination, innovation, artistic ability, cooperation and student leadership skills were given equal value as computing, reading and writing in evaluation. Every two-part summative portfolio-style assessment included an oral presentation of relevant BIG IDEAS that respond to unit essential questions (EQ) and a model of either the problem or solution as informed by students' research and learning activities.

Students were asked to read/view relevant texts (teacher-selected and student-selected) to support their learning and perform appropriate analyses of these texts to support their understanding. Evaluation of each unit was based on assessments designed by the teacher and products generated by students.

Teaching and Learning

As organizing principles for evaluating my teaching and the student learning it produced, technology and engineering products were the primary means of assessment. Although pre- and post- diagnostic surveys were administered to students, grades were determined solely on students' ability to demonstrate their own growth. This offered students a unique opportunity to demonstrate what they learned, to interact with stakeholders within and outside of school while preparing themselves for varied and multiple applications of science and mathematics in a real-world work context so "…that the learning experience be a meaningful one through which the student [could] acquire some of the instruments and scientific-technological modes of thinking."[16] To that end, a focus on teaching goals (curricular objectives) and learning goals (student learning outcomes) consistent with the Ilan (*ibid.*) model is outlined below. Ilan's work offered particularly salient perspective because it was part of a larger workshop to bust eight myths in STEM education (miscellany, STEM knowledge linearity, coverage, detachment, critical thinking, a singular scientific method, universal utility and curricular homogeneity) that sustain exclusion of some groups and in many ways contribute to oppressive practices. These myths were exposed by the keynote speaker[17] as what can only be imagined to be a jolting opportunity to recharge STEM education praxis.[18] In the March 2000 address, Osborn challenges participants to think about this truth: "As currently practiced, science education rests on a set of arcane cultural norms." In that moment, Beijing became a center of incredibly transgressive thinking about STEM education. As participants exposed their willingness to challenge formerly "unquestioned norms" of STEM education, they established an agenda for STEM education equity that is rarely seen. Taking on the need for curriculum reform, assessment reform and stakeholder engagement, an agenda for 21st century STEM was established. From that vision, I wrote a curriculum for a new course whose teaching and learning goals would honor the clarion call made by that group of progressive thinkers. Table 7.2 outlines those goals.

Table 7.2: STEM-21 Teaching and Learning Goals.

Teaching goals	Learning goals *(Students will be able to ...)*
(1) Develop student recognition of discovery, science and technology as part of human enterprise and social/societal structure (2) Develop curiosity and interest in science, technology and related topics that have implications for public debate (3) Develop an understanding about the cumulative nature of public knowledge (4) Develop an understanding about the interdependence of humanity and the environment (5) Foster a spirit of intellectual integrity and respect in an increasingly more participatory culture & social structure	(1) Provide scientifically sound explanations of various phenomena based on available knowledge and technologies (2) Recognize and execute various thinking strategies that are characteristic to science and technology (3) Read, summarize and critique popular science periodicals (4) Describe and organize the historical course of scientific thoughts (5) Work independently and collaboratively to demonstrate understanding and support opinion in written and oral form

Source: Author.

Science and mathematics anchor instruction in STEM-21. All units of instruction present physical science and mathematics content. Student learning outcomes were measured by students' ability to model conceptual understanding using digital (technology) or physical (engineering) means. The concept of technology, defined as any type of human-made system or process (NRC, 2011), meshes well with the idea that the designed world must adhere to constraints established by experimentation, testing and materials (engineering). The goal of the STEM-21 class is to build student competency and technical literacy in each of the following areas:

- Collection, evaluation and representation of real-world data
- Creation of models to accurately represent phenomena observed in the physical world
- Communication of complex ideas in quantitative and qualitative ways
- Understanding of 21st century pre-professional and training pathways to STEM careers and responsible citizenship

Portfolio Assessment

The portfolio is a collection of learning episodes for students compiled over the course of the unit. Assessment is designed to be presented to an authentic audience of stakeholders. In addition to science, technology, engineering and mathematics components, aspects of the human experience and human systems that shape our need for innovation and creative solutions to problems will be discussed. The unique sociological characteristics of the 21st century were compared to previous generations in an effort to imagine life in future generations. Students created portfolios for each unit of instruction that included samples of each type of product generated during the unit. Content quizzes (traditional & online) based on science and math concepts, independent practice assignments, guided practice assignments (classwork), inquiry activities (lab work), next-step questions, and model-planning sheets, each providing evidence[19] of independent thinking and effort. Students were expected to collaborate in work groups for each unit of instruction. These work groups, responsible for small group tasks (for example, lab activities, group share-out), required cooperation between students. Students must get feedback from a member within the group as well as an individual outside of the work group during round-table discussions held before performance/product submission. Each portfolio must contain a checklist and was evaluated according to a performance rubric based on five domains of evaluation: performance/products (models), effort, achievement, submission standards and questions. I call this the PEAS & Qs performance rubric (Table 7.3).

The name (PEAS & Qs) is itself an acronym designed to tear down two often unintended barriers to student metacognitive fortitude: tiered vocabulary[20] and coded language.[21] The use of vernacular speech and idiomatic phrases keeps students outside of the established norms of dominant culture and tends to amplify uncertainty and timidity in the classroom. Having witnessed many instances when students "don't get it," and consequently choose not to

Table 7.3: PEAS & Qs Portfolio Rubric.

Measure	Description
Performance/products	Models created by the student to demonstrate content understanding that clearly represent the major ideas inherent in the discipline and meet the goals/objectives outlined in the course scope and sequence and/or curriculum
Effort	Student works to become increasingly independent and interdependent (rather than co-dependent and/or fully dependent on peers or teachers) in approach to content thinking and knowledge construction; closely related to initiative, effort implies that the student has considered the intended audience for the model/representation in a way that is appropriate and focused
Achievement	Student clearly meets the goals and objectives of the assignment/task while demonstrating understanding of the content from a variety of perspectives
Submission standards	Evidence of purposeful selection and organization of artifacts on part of the student author is apparent; product is submitted on time, submitted using an appropriate format and includes checklist for the assignment
Questions	Questions posed to scaffold previous learning, available information and next steps in research are original and demonstrate sophisticated thinking patterns; these questions must confirm, extend or attempt to refine the work/observations made by others as cited in previously published works

Source: Author

participate, there is a particularly disheartening regret that I have when I think about students I have served with any kind of communication dis-ability. By providing opportunities for students to use coded language that is unique and/or created within the context of the small community of a classroom, a level of membership is created between students that celebrates the shared and unique experiences they have, knitting them together and building their identity while also clearly communicating big ideas beyond and including content, like safety, security, the shared responsibility of STEM education for better local and global citizenship. Finally, empowering students with their own tools for self reflection is a means of providing them agency to understand their shortcomings and their achievements and develop high expectations for all stakeholders around them, including the teachers who guide them.

Stage 3: Conceptual Framework for Action Research

The National Science Teachers Association (NSTA) offers position statements[22] on various topics of interest to teachers attempting to become transgressive practitioners. Discussions of multicultural science education, gender equity, English language learners, students with exceptionalities and teaching science for societal and personal meaning are among these positions. All fit within an equity framework for STEM education that when applied would be a foundation for transgressive practice[23] however the discussions themselves seem prescriptive, especially to the novice teacher. With this framework in mind, I consistently interrogate my own praxis around three questions:

1. How can I impact students' understanding so that they can:
   a. Explain physical science principles using appropriate vocabulary?
   b. Describe phenomena about the material world based on evidence collected during lab?
   c. Contribute to larger discourses about the material world?
2. Which methods for communicating technical information work best with adolescents for knowledge building across multiple diversity texts?
3. Which curricular questions help students to understand the sociological limits to participation of all communities in STEM at the citizen and/or professional level?

Stage 4: Sample Lessons in STEM Disciplines with Reading and Art in Mind (STREAM).

Between 2012 and 2016, the years when I taught the course, the content of the course had changed only slightly with the exception of one thing: an increased intentionality around literacy. The changing class/student profile in 2016–2017 to include multilingual students and/or students with communication or language processing exceptionalities necessitated a mindful shift in focus in instruction and assessment. Inspired by previous work done with Kara Viesca of the University of Nebraska, communication was prioritized. This way of crafting lessons helped me to to understand that students with limited English proficiency or language processing exceptionalities have an embedded tool for being part of the class community, work also being advanced by scholars like Ana Maria Villegas, Ohkee Lee and others. Language was thus used as a leverage point with youth: a place of entry and a tool for participation in inquiry activities. At a week-13 check point, students were asked several questions about the course including how often they read STEM-related content in and outside of class, their favorite STEM-21 discipline and their preference for inquiry activities (pre-scripted protocols or discovery-based) and technology integration (with mobile phones, school-provided tablets or laptops). Their responses helped me to understand how to shape the course for the last thirteen weeks.

Sharing resources is itself a transgressive practice; out of fear of copyright infringement or distrusts surrounding crediting, many are reluctant to share in open forums. I have adopted a different understanding. Instead, I embrace a principle of *confirmation, extension and/or refinement of others' work* (C.E.R.O.W. or ZERO) as the new standard for knowledge construction; without these purposes, any community trying to build new insights has nothing. Here, I take two approaches to that. The original lesson plan outline[24] for the STEM-21 class is too substantial for sharing here; it is provided in Appendix V. By sharing a topical view of the first month of instruction (Table 7.4) however, I provide insight into the first weeks of instruction.

Table 7.4: Lesson Plan BIG IDEAS: Words Tell the STEM Story (Month View).

Monday	Tuesday	Wednesday	Thursday	Friday
			9/1	9/2
9/5 Labor Day	9/6	9/7 (D) Introductions, Expectations and Class Norms	9/8 (D) Shared Science Interviews & Video	9/9 (D) Designing for Disaster One Word at a Time: Shattered
9/12 (A) Science	9/13 (B) Technology	9/14 (C) LAB—Engineering for Real Problems (Water holder: Design a diaper)	9/15 (D) Mathematics, Everyday & Scale "Math for the Win"	9/16 (E) 21st Century *Current Events: STEM Business Challenges*
9/19 (F) Society *PBS Election, Introduction* 50 for 50: Letter to the Next President Sample-Food Waste	9/20 (G) LAB—Engineering Materials *Textile study*	9/21 (A) Literacy: Stuff Matters (Audio)	9/22 (B) Stuff Matters (continued)	9/23 (C) LAB—Engineering for Retail (Pack & Ship Chip)
9/26 (D) Mathematics, Measurement	9/27 (E) Stuff Matters: Indomitable. *Metal properties study*	9/28 (F) Modern. Discussion of the "Ages"	9/29 (G) Technology Tool LAB	9/30

Source: Author.

The essence of the STEM-21 model is captured in this concentrated view. Purposefully designed to place societal conflict at the center of all discussions about scientific, technological, design/engineering and quantitative reasoning related to human industry and enterprise, the first week sets a tone for considering why and how STEM serves important functions in service to society. Our place, only fifty miles away from "Ground Zero" makes teaching in this way particularly relevant since our community suffered significant and deeply personal losses on September 11, 2001. Choosing to ground all STEM study in this context allowed class stakeholders to understand a broad range of content, across multiple disciplines, making sense of racial and cultural tension and contextualizing what have become post-millennial[25] ways of thinking, living, and being.

Week Two: STEM in a 21st Century Society Defined

In terms of instruction, I wanted to make sure that the rationale for approaching STEM in a 21st century societal context was clear to students—disrupting notions of domination and placing STEM in human context.[26] Students enrolled in the course in 2016 represented the first class of students born too near to 2001 to perhaps understand those losses by first-hand experience;[27] this shift in direct experience represents a change in generation. Week two study began on a Monday. The preceding lesson on Friday was based on a Times magazine photo-essay[28] of the tragic losses of September 11th 2001 called "Shattered". Although the photo-essay has changed over time, the images served as a point of "vicarious observation" of problems that have natural science and social science roots. Base-level observations of dust and smoke or corrosion were entry points into conversation about environmental problems while images of grief and desolation represented more intricate and higher order thinking.

Defining STEM in a 21st century societal context meant deconstructing each domain (science, technology, engineering, mathematics, 21st century timing and societal issues) first. Using a classroom learning management system, presentation slides were shared and updated in real-time to incorporate the dynamics of class discussion. Finalized week two slides are presented here.

Monday's presentation highlighted the multilingual composition of our class. Enrolled in the class were students whose primary home communication language was English, Taiwanese, Japanese, French, Spanish, Creole and with assistive technologies. Day one was a presentation of the scientific method and orientation to the science practices outlined in the NGSS. Day two instruction was based on technology tools, the concept of invention and the resistance to the use of "good and bad" as norms of communication in our class. Very early in the class I wanted students to know that values are assigned to objects based on their usefulness at the time. This discussion of inventions[29] allowed us to fully consider why technology

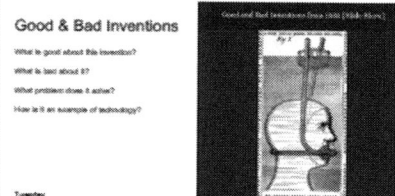
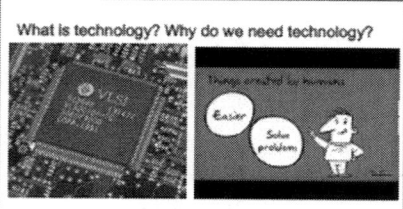

Figure 7.1: Defining S & T in STEM: STEM-21 Teaching Slides. Source: Author.

exists in context of people who do work. Day three was an exploration of engineering and the design process; from a practical perspective, this was the "lab" day in the cycle. In our school, every seven-day cycle within lab courses has two dedicated block periods (94-minutes of direct contact).

Day four was our mathematics orientation day. Building on the need to expose students to an authentic audience and share examples of desired products, students were introduced to a calculator company contest that asked for shared photographs on everyday use of math. I shared a photograph I had taken of a water fountain in

ENGAGED INTERDISCIPLINARY LITERACY | 181

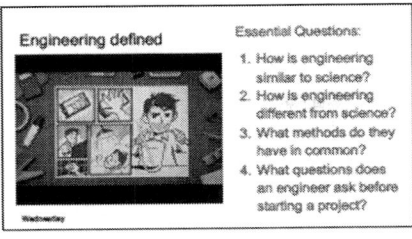
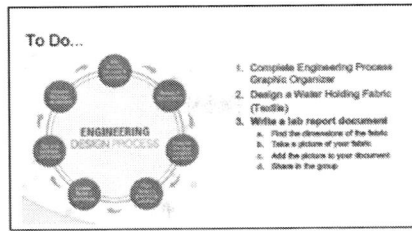

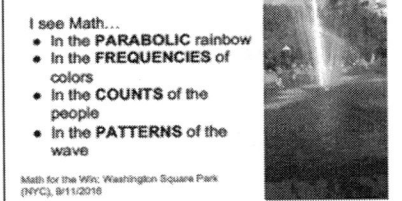
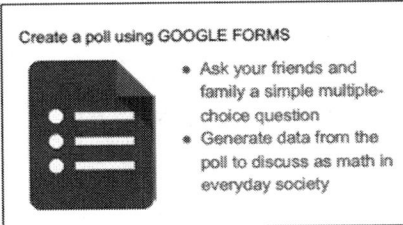
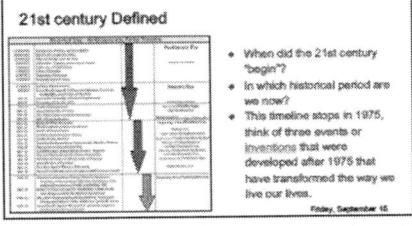

Figure 7.2: Defining E & M in Society Context: STEM-21 Teaching Slides.
Source: Author.

a park describing parabolic path and properties of light (the rainbow) as an algebraic model of frequency and wavelength. Day five was a presentation of the 21st century as a construct of time. We explored the theoretical underpinnings of "age" in terms of knowledge construction and social thought. Day six was a presentation of the sociological implications of STEM.

Placing STEM in direct conversation with societal challenges around healthcare was critical. Because pharmaceutical science is a major industry in New Jersey, students could connect directly with the discourse around drug prices, public

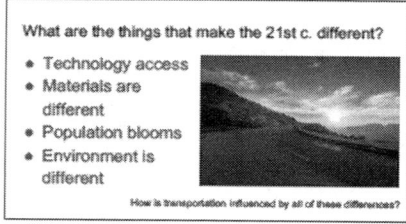

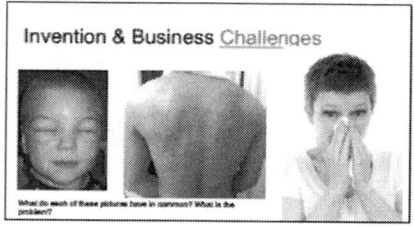
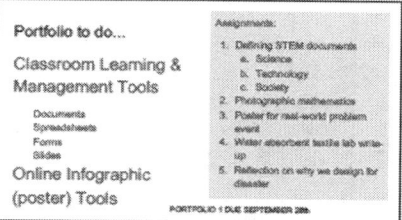

Figure 7.3: Defining a Role for STEM in the 21st Century Society: STEM-21 Teaching Slides. Source: Author.

health and innovation (research and development). In our corridor of the state, several large-scale and small-scale companies employed community members in professional and non-professional ways as paid and volunteer agents. This discussion allowed for differentiation around reading level.

Recommendations for Differentiated Assessment

The challenge of differentiated instruction is not really about designing different activities for students to do. The real challenge is how do we differentiate assessment. Even though student-produced evidence of learning is different and perhaps distinguished by student interest, student ability, language/literacy level and course entry performance level transgressing beyond "fairness" toward "equitable" is an imperative that needs to be addressed. To meet this challenge, there are a few things that can be done:

1. Create and share rubrics BEFORE instruction
2. Provide students with templates or models of expectations (i.e., sample products)
3. Select content that is of high-interest to the students being served in the class (which ultimately means we change the instructional material every year/semester)
4. Revise and re-administer assessments if student results suggest a low level of understanding; embedded in this requirement is the need to reteach content without complaint
5. Survey students often to determine interest and modify plans based on feedback

In the STEM-21 course, students were encouraged to read scholarly peer-reviewed articles[30] on the subject (high level), read daily periodicals like the NY Times[31] "Science Times" articles on the subject (intermediate level) or watch and analyze sections of a close-captioned discussion[32] on the subject (basic level). We concluded the week on day seven by contextualizing all of it (science, technology, engineering design, mathematics, time, societal needs and the sociopolitical process). Our final exercise for the week was to review expectations for evaluation. We discussed the parameters of the portfolio, had opportunities to use various tools to communicate our understanding and judge each other's work as critical friends.

In the fall of 2016, the political climate surrounding the presidential election fueled tensions in the community. The small population of Black and Brown students was not always safe to openly discuss their thoughts, mainly because they really are a minority in this school. Except in this class. Being taught by an equity-minded teacher who has navigated similar academic terrain as a Black student and educator became critically important. Building on principles of rigor with relationship, relevance with fun, we created a safe space to challenge the status quo and enjoy learning together. The final attribute of Stage four lesson study involves reflection and analysis with others. As the course evolves, new topics for STEM-21 are emerging. The discussion questions that follow invite extended conversation about topics that researchers and practitioners could both explore (separately and as a collective) to develop meaningful platforms for equitable STEM teaching and learning going forward. In posing these questions, I am attempting to be transparent about the adaptive nature of curriculum writing, hoping that many will take up this cause, pose new questions and present new lesson ideas that expand the borders of STEM and encourage engagement by most.

Discussion Questions

- What topics would you use to construct a STEM-21 Unit for your population?
- Using social media or listserv chat space, identify a line of STEM-21 discourse worth exploring with students.
    a. What are the most common hashtags in that space?
    b. What media/technologies are used to stimulate discussion?
- Create a photo-essay to document a recent disaster (e.g., California forest fires, Hurricane Maria of 2017). Invite conversations about the STEM-21 implications for community response to the problems created by these disasters.
    a. As youth still defining your life path, what problems would you want to solve as a result of these disasters?
    b. What roles do you imagine for everyday citizens to secure safe living conditions for communities under these circumstances?

Notes

1. Implications of Julie Brown's study highlight the need for pre- and in-service teachers to have clear examples from research about instruction they are expected to provide. Action research and lesson study designs provide that insight. See Brown, 2017.
2. Osborn (2000, p. 11) presents several key arguments about how a limited perspective on STEM as a "linear progression" of the work of a few good men leads to STEM education that is dogmatic and naïve.
3. Miles & Darling-Hammond, 1998.
4. Based on 2010 census data available online at https://www.census.gov/quickfacts/NJ
5. Available online at http://www.princetonk12.org/district_content/About_PPS/Mission
6. See "A Guide to Disability Rights" at https://www.ada.gov/cguide.htm#anchor65310 for additional descriptions.
7. Doctoral study had prepared me to understand current STEM education trends and teacher education policy reforms, including the National Research Council Framework which eventually became a foundation for NGSS.
8. In 2011–2012 school year, students were also able to take AP Environmental Science if they met the grade completion criteria (B+ or higher in both biology and chemistry).

9. As part of the planning process, rubrics and sample products were created by the teachers. The intent was to be transparent with students about expectations based on knowledge of the requirements to use innovative tools. We never expected students to do what we had not done and yet we knew that they would probably exceed us in their products based on their comfortable and familiar use of technology.
10. See McGraw Hill webinar online at https://www.youtube.com/watch?v=aRe-cpt799E&index=58&list=PLYmYDLb2oJqEzVcH_VQdAkDUv7M_ofm2I (January 22, 2013): "Chemistry and the Next Generation Science Standards"; three teachers within the department share resources at three levels of student performance.
11. See Appendix D Case Studies at https://www.nextgenscience.org/appendix-d-case-studies
12. Committee on Conceptual Framework for the New K-12 Science Education Standards; National Research Council, 2012.
13. As an early adopter of NGSS, the New Jersey Department of Education developed model curricula, including a capstone course to guide a variety of education entities in thinking about the new standards. See http://www.nj.gov/education/modelcurriculum/sci/capstone.shtml for additional details.
14. See "Defining the generations" https://uk.sagepub.com/sites/default/files/upm-binaries/58778_Abrams_The_Multigenerational_Workplace_Ch_1.pdf and "Millennials in the Workforce" (video https://www.youtube.com/watch?v=QXWNChoIluo&t=615s) by Simon Sinek for more insight.
15. Blaschke, 2012.
16. Ilan, 2001, p. 116.
17. Osborn, 2000.
18. The term praxis is applied here to distinguish actual actions taken by teachers as they implement theory, (e.g., putting theories about teaching and learning into practice in the classroom).
19. A claims-evidence-reasoning approach to evaluation was taken as a mechanism of differentiation for individualized learning.
20. The National Clearinghouse for English Language Acquisition publishes resources that share research and practice that address levels of vocabulary in vocabulary instruction. The Summer 2011 quarterly review looks specifically at STEM for English Language Learners. Available online at http://ngl.cengage.com/assets/downloads/ngsci_pro0000000028/ngsci_research_accellerate_v3i4_062011.pdf
21. The emergence of memes in social media provides an example of high-context coded language that serves as a port of sorts to specific cultures. Implicit and explicit meanings that are understood at various levels by a highly prescriptive or selective group of people can be interpreted in memes, pushing people or keeping people out of conversations.
22. A full list of "position statements" is available online at http://www.nsta.org/about/positions/#list
23. Transgressive practices were characterized by bell hooks as those activities enacted by educators (and students) that courageously go beyond boundaries in *Teaching to Transgress*. (1994). New York: Routledge. Sharing success and failure is transgressive.
24. Appendix X contains a full description of the STEM-21 class.

25. See Michael Dimock's discussion "Defining generations" at http://www.pewresearch.org/fact-tank/2018/03/01/defining-generations-where-millennials-end-and-post-millennials-begin/
26. This notion fits well within the equity framework developed by Gutiérrez (2002) that challenges inequity while also attempting to bring into alignment pedagogies of dominance and critical (i.e., critical race, critical language) importance.
27. It is worth noting that there were members from within this NJ community deeply impacted at a personal level by these tragedies. Careful and thoughtful community, district, school-wide and classroom level activities honor the loss of lives and shifts in cultural norms related to 9–11.
28. The "Shattered" photo essay (http://content.time.com/time/photogallery/0,29307,1660644,00.html) has changed over time and is used to ground discussions about how art and history can be used as means to convey real-world problems and solutions in STEM.
29. Scientific American published slideshow traces the history of invention is available online at https://www.scientificamerican.com/article/good-and-bad-inventions-from-1865-slide-show/
30. Allen, 2016; Lexile (reading) level: 1,560 is appropriate for college-ready seniors.
31. Rosenthal, 2016.
32. Prescription Drug Market Hearing https://www.c-span.org/video/?404183-1/hearing-prescription-drug-market

References

Allen, L. (2016, May 20). Prescription drug costs: Do research expenses justify eye-popping prices? *CQ Researcher, 26*(20), 457–480.

Blaschke, L. M. (2012). Heutagogy and lifelong learning: A review of heutagogical practice and self-determined learning. *The International Review of Research in Open and Distributed Learning, 13*(1), 56–71. Retrieved from http://www.irrodl.org/index.php/irrodl/article/view/1076/2087

Brown, J. C. (2017). A metasynthesis of the complementarity of culturally responsive and inquiry-based science education in K-12 settings: Implications for advancing equitable science teaching and learning. *Journal of Research in Science Teaching, 54*(9), 1143–1173.

Gutiérrez, R. (2002). Enabling the practice of mathematics teachers in context: Toward a new equity research agenda. *Mathematical Thinking and Learning, 4*(2–3), 145–187.

Ilan, M. (2001). Designing an interdisciplinary curriculum in science and technology. In M. Poisson (Ed.), *Science Education for the Contemporary Society: Problems, Issues and Dilemmas—Final report of the international workshop on the reform in the teaching of science and technology at primary and secondary level in Asia: Comparative references to Europe* (pp. 111–118). Retrieved April 17, 2012 from http://www.ibe.unesco.org/fileadmin/user_upload/archive/curriculum/China/Pdf/beijingrep.pdf

Lewis, C., Perry, R., & Marata, A. (2006). How should research contribute to instructional improvement? The case of lesson study. *Educational Researcher, 35*(3), 3–14.

Miles, K. H., & Darling-Hammond, L. (1998). Rethinking the allocation of teaching resources: Some lessons from high-performing schools. *Educational Evaluation and Policy Analysis, 20*(1), 9–29.

National Research Council. (2012). *Framework for K12 science education.* Retrieved from http://www.nap.edu/catalog.php?record_id=13165

Osborn, J. (2000). Keynote speech. In M. Poisson (Ed.), *Science Education for the Contemporary Society: Problems, Issues and Dilemmas—Final report of the international workshop on the reform in the teaching of science and technology at primary and secondary level in Asia: Comparative references to Europe* (pp. 8–14). Retrieved from http://www.ibe.unesco.org/fileadmin/user_upload/archive/curriculum/China/Pdf/beijingrep.pdf

Rosenthal, E. (2016, September 2). *The Lesson of Epi-pens.* Retrieved from https://www.nytimes.com/2016/09/04/opinion/sunday/the-lesson-of-epipens-why-drug-prices-spike-again-and-again.html?_r=0

CHAPTER EIGHT

Transformative Education Pathways to Improve Health Literacy, STEM Learning, and Youth Outcomes

GRETCHEN E. L. SUESS,[1] JOANNA CHAE,[2] AND SHARON LEWIS[3]

Abstract

This chapter shares findings from a mixed-method study using data from the 2010–2016 Health Sciences Education Pipeline Program (Pipeline) at the University of Pennsylvania (Penn). Pipeline is a partnership between the Perelman School of Medicine, Netter Center for Community partnerships, and two public high schools (William L. Sayre and West Philadelphia High School.) The

---

1 Gretchen E. L. Suess, Director of Evaluation
  Netter Center for Community Partnerships & Department of Anthropology
  University of Pennsylvania
  gsuess@sas.upenn.edu
2 Joanna Chae, Director of Moelis Access Science (August 2013–December 2017)
  Netter Center for Community Partnerships
  University of Pennsylvania
  cjjc2274@columbia.edu
3 Sharon Lewis, Clinical Assistant Professor of Neurology
  Perelman School of Medicine
  University of Pennsylvania
  sharon.lewis@uphs.upenn.edu

primary goals of Pipeline are to: (1) Enrich STEM teaching and learning by supporting high school teachers with expanded physical and life sciences curricula; (2) Educate high school students about fundamental principles of medicine and diseases that affect their community while improving communication and cultural competency among all participants; and (3) Improve health care and health literacy by increasing underrepresented student participation in health-based careers and post-secondary education.

Editorial Reflections

I, Joy, "met" W.E.B. DuBois when I was a junior in high school. That was the first time I had ever heard any reference to the "Philadelphia Negro" or the "talented tenth." I had never known that Black people were so studied but Philadelphia was different. At that point in my life, my family was proud of my "good grades" and wanted me to try to forget about their stories—but theirs were my own.

My grandfather was born and raised in Alabama in the early 1920s. He had a seventh grade education and spent most of his time as an apprentice in a mechanical shop outside of school. He matured and eventually became a linesman mechanic at Ford. He was a major influence on my life. As a college graduation present, he took me to his hometown in Alabama. By then I was old enough to know about the legacy of exploitation of Black people in health, medicine and the STEM research community. Details about the mistreatment of the Tuskegee Airmen were as troubling as the Tuskegee syphilis experiments. Until the day he died, my grandfather never trusted Western medicine administered by White doctors, even though his wife, my grandmother, was a nurse. He died of cancer having accepted very little treatment—all he expected in life was his three score and ten based on the scriptures:[1]

> *The days of our years are threescore years and ten; and if by reason of strength they be fourscore years, yet is their strength labour and sorrow; for it is soon cut off, and we fly away.*

I am not sure why but the problem of medical mis-/dis- trust has persisted through the generation of my fathers (my birth father dying violently and suffering from a lack of treatment for mental health "unwellness;" my stepfather died from a rare cancer that could have been treated—he watched a cyst grow from a pea-sized "clip" into a grapefruit-sized tumor before seeking medical attention)[2] *in spite of better educational attainment. There is a public health crisis in securing the life of Black people, especially Black men, to the projected American life expectancy of almost four score. When reading the upcoming chapter on "Transformative Education Pathways to Improve Health," my initial response was as a teacher: how should STEM educators mitigate and/or mediate the lack of trust that exists between communities of color and the healthcare system, especially in cities? My more visceral response was to the interaction of poverty, stretched resources, public health crises, and human diversity; these are overwhelming structures cemented as institutional memory in organizations and people. These are systems based in power.[3] If this lack of trust had stopped in my grandfather's generation, would I feel the same way? I wondered as I read if STEM education would ever be enough.*

Far away from cities, I, Janelle, did my dissertation work in mostly rural, Indigenous communities of Mexico and Guatemala. In a sickening link to Joy's story, Guatemala was also a site for U.S. sponsored syphilis experiments. My research examined the international development programs of U.S. based organizations' professional development for teachers, centering the perspectives of the teachers in these communities. When I first realized what I wanted to study, I had trouble finding related research. I didn't know what terminology would help me search the literature. I consulted with the research specialists at the university library—they were unable to distinguish the name of the field either. I was frustrated since all I could find were reports to the organizations' funders of what they had done; I couldn't find research that asked the community members what their perspective was. There were also chapters by researchers in various fields, including education, who described their work with teachers in so-called developing countries, also essentially descriptive of what THEY had done. But there was no terminology to capture the transnational nature of work, or the power differentials between the privileged foreigners and the locals; there was little attention to the communities' histories, or cultural beliefs, or marginalization within the nation-state. As I did my research, it became clear that all the development organizations framed their work as "well intentioned." Their outsiders' metrics calculated need in these communities, but most of the work did not incorporate local voices or Indigenous knowledge(s) into their frameworks.

The upcoming chapter helps us to know that there is still work to be done. Our proposed solutions still need to be scrutinized and our models will still benefit from cast doubt. As editors we remain hopeful that as more of us find the courage to be reflexive about social justice and public health, we will erect fewer monuments to our good works and incremental or temporal victories. The need to continue the work of going back into communities where we seem under-appreciated is great but we owe it to the communities we serve to be there.

Introduction

While the general health of Americans has improved according to national level statistics, disparities continue to persist. Racially and ethnically marginalized populations have more limited access to care and worse outcomes than their White counterparts. According to the National Center for Health Statistics in 1950, there was a nine-year difference in life expectancy between Whites (69.1 years) and Blacks (60.8 years). This difference in life expectancy continues today. To challenge this problem, we must engage in collaborative critical reflection about the social systems that perpetuate this reality and what we can do together as socially-engaged scholars, educators, and practitioners to dismantle them. Data have shown that diversifying the health education environment and future health care workforce who provides care to marginalized populations have direct impacts on reducing the health care gap. Specifically, a thorough review of existing research demonstrates how patients from nondominant communities are more likely to experience longer care visits, increased satisfaction with their care, and will more likely follow their physician's advice if race, ethnicity, and/or language concordance with their physician is present.[4] In addition, early and continuous health care discussions can increase levels of health literacy to further counter this gap and reduce associated risks of poorer health outcomes and poorer use of health care services.[5]

Yet, diversity among the U.S. physician workforce has not changed significantly over the past few years. Even though Blacks comprise the second largest racial group in the US, they account for only 4% of the physician workforce.[6] These racial and socioeconomic disparities persist for marginalized populations due to the limited access to medical and health-based educational and career pathways. The Association of American Medical Colleges (2016) reported that the percentage of medical school applicants who are African American, Latinx, and Native American, which combined represent less than 15% of applicants, are disproportionately below what would be expected given their representation in the population. Black, Latinx, and low-income students are also generally underrepresented across all postsecondary enrollment despite the national growth in the last few decades.[7] More specifically, STEM achievements also directly impact access to professional medical fields and while the types of degrees granted and level of student interest in STEM has increased since 2000, there remains a significant gap in gender and between Whites and non-Whites. The 2016 U.S. News/Raytheon STEM Index shows that between 2011 and 2016 while the number of White students who earned STEM degrees grew 15%, the number of Black students fell by roughly the same margin.[8]

The Obama-era Every Student Succeeds Act (ESSA) of 2015 and the Next Generation Science Standards (NGSS) in life and physical science call for innovative evidence-driven programs to improve secondary education, increase STEM teacher competencies, increase the acquisition of core science content knowledge, and the direct application of those concepts. In order to expand STEM experiences, learning, and career opportunities to nondominant youth they need to be provided with culturally-relevant science opportunities that allow them to engage in "productive struggle" as they tackle problem-solving and authentic applications of science knowledge.[9] The Health Sciences Education Pipeline Program (Pipeline) at the University of Pennsylvania serves as a potential model for embracing these ideals and illustrates the promise of a place-based approach to improve STEM teaching and learning in the physical and life sciences, increase underrepresented student participation in health-based careers and post-secondary education, and engage high school students to deepen their understanding of the fundamental principles of medicine, diseases that affect their community, and health literacy. This chapter shares findings from a mixed-method study of the Pipeline Program.

Methods

Research and evaluation on this work are ongoing, led by the Netter Center's Director of Evaluation. This study draws from both quantitative and qualitative data for high school seniors who had been in the Pipeline Program between spring 2011 and fall 2016, with additional enrollment data for three other core Netter Center programs (Agatston Urban Nutrition Initiative; College Access and Career Readiness; and Extended Learning Afterschool and Summer) at two high schools in West Philadelphia. Specifically, data include high school enrollment records for 1,533 seniors between 2010–2016 and GPA between 2011–2014 provided by the School District of Philadelphia, program administrative data such as participation and attendance between 2010–2016 from the Netter Center,[10] college matriculation records for every term beginning with summer 2010 through fall 2016 from the National Student Clearinghouse StudentTracker,[11] semester pre/post surveys from 118 high school students over six years, and focus groups with 26 high school students in the Spring of 2011, 2015 and 2016. Quantitative data were analyzed using SPSS and R. Statistical methods include principal component analysis, multiple regressions, t-tests, and analysis of variance (ANOVA). Qualitative data were analyzed for themes regarding student engagement, implementation, and impacts in STEM learning using Atlas.ti.

Background of the Pipeline Program

The Pipeline Program was launched in 1998 and engages 9th–12th graders at two neighborhood comprehensive public high schools located within the University's local geographic community (West Philadelphia and Sayre High Schools), where approximately 37% of local youth live below the poverty line.[12] Both schools are predominantly African American/Black, making up between 96–97% of the student body.[13] In 2015, both schools also had significantly lower college matriculation rates than the 60% overall school district rate.[14]

Hill (2008) claims that "disadvantaged students and their families tend to be more dependent on their schools for access to resources that are related to postsecondary educational attainment" (p. 56). In line with this, policy recommendations for improving college readiness have included proposals to raise standards required for high school graduation, improve assessment measures used in exit-examinations, promote college going cultures in high schools, and establish more sophisticated data-tracking systems to monitor student progress and flag those at risk of not graduating on time.[15] However, another argument posits that college readiness is more than just course taking and standardized test scores; schools need to stress four indicators for college readiness: content knowledge and basic skills, core academic skills, non-cognitive behavioral skills, and college knowledge.[16] Over the past 20 years, a number of creative strategies have been designed and implemented in an effort to improve college readiness, access, and enrollment. However, relatively few of these programs have been rigorously researched or evaluated for overall effectiveness and those that have been studied have generally been found to have modest to no impact.[17] Moreover, often, such programs have targeted the more advantaged segments of underserved populations who need fewer supports and are identified as being most likely to succeed.

Since the late 1980s, Netter Center leaders and their Penn colleagues have been implementing programs that test the proposition that if colleges and universities can succeed in transforming themselves into genuinely engaged civic institutions, then they will be better able to achieve their self-professed, historic missions of advancing, preserving, and transmitting knowledge; and they will reduce gaps in opportunity and achievement. The work emphasizes that such partnerships must be mutually beneficial to all participants and have positive impacts on both high school and Penn students to transform institutions.[18] Advocates for the full participation of multiple communities in universities call for the integration of projects and people driven by a shared vision of equity, diversity, and inclusion with regards to the community, public, and civic work of universities.[19] The Pipeline Program is explicitly focused on neighborhood comprehensive public high schools near the University campus to embrace three core goals of the program: (1) Enrich STEM teaching and learning by supporting school day teachers with an expanded curriculum in the physical and life sciences; (2) Educate high school students about fundamental principles of medicine and diseases that affect their community while improving communication and cultural competency among all participants; and (3) Improve health care and health literacy by increasing underrepresented student participation in health-based careers and post-secondary education.

The program uses a place-based approach with neighborhood public schools to improve college access and secondary education opportunities that does not involve *creaming* top students, while simultaneously improving undergraduate and graduate-level medical education. The Pipeline Program recruits Penn undergraduates, first- and fourth-year medical students, medical residents, and medical faculty from Perelman School of Medicine every year to coordinate and implement the program with high school physical science, chemistry, and biology teachers to explore topics such as neurology, cardiology, gastrointestinology, infectious disease, and veterinary medicine. This multi-tiered approach exposes high school students to the application and translation of core life science concepts and new opportunities for medical and education careers. At the conclusion of the program, students create a final presentation to bring healthcare awareness to their community.

Fostering a Science Identity and School Engagement

The success of the Pipeline Program is based, in part, on the use of pedagogical strategies that enhance learning through the exploration of diseases that affect student communities, whereby students also learn the fundamental principles of medicine and deepen their ability to apply science knowledge. In doing so, youth also begin to formulate their science identity. Science identity is one's sense of self, capabilities, and place in the world concerning what one wants to do and become in regard to science.[20] Science curricula that students find to be relevant to their lives are shown to improve student engagement. Such curricula also recognize that students bring a wealth of science-related, life experiences, and build upon their knowledge.[21] These lessons can also bridge the gap that students may have established between science and their lives, which helps students see the practicality and meaning of science.[22] This is a critical need when many students, along with their families, consider math and science irrelevant to their personal interests, goals, or future career prospects when more jobs today require that knowledge.[23]

Following participation in the Pipeline Program, students highlighted a sense of deepened science identity and understanding of how science directly informs their lived experiences in the Spring of 2011 and 2016: "I learned that a lot of stuff that we learned in the Pipeline Program ran through my family a lot, so it was a good thing to learn … how it affects you and what age group you get it." A different respondent stated: "My sister has seizures a lot and I never knew what to do, so I took [the lesson] to heart a little bit. Now I know what to do. I feel more helpful like I can do something." A third respondent said: "[We] talked about HIV and all the different spots it affected ... I know like one or two people that have HIV and I always wondered why they act like that."

Our findings reveal that increased participation in the Pipeline Program over the years was strongly positively correlated with students reporting that "science is interesting" and beliefs that "scientific discoveries have an impact on everyday life" ($p < .001$). In addition, students who participated in multiple semesters of the Pipeline Program expressed higher levels of being interested in their high school education and continuing their education after high school ($p < .01$). Likewise, students in the graduating Class of 2015 who had participated in the Pipeline Program maintained their GPA of 2.5 over three academic years while their peers experienced a significant and steady decline from a 2.7 in 2011–2012 to a 2.5 in 2013–2014 ($p < .001$, $p < .05$, respectively).

The Learning Environment, STEM Mentorship and Unpacking Cultural Competence

One factor that directly shapes the ways in which low performing students learn is the way in which they are perceived by educators. Subject-formation and self-identity go hand-in-hand. As teachers are trained to serve in low-income schools, their conceptions about poverty and academic capabilities of urban high school students transfer directly into teaching practices.[24] Engaging youth with culturally-relevant science opportunities and application of new knowledge is also not always possible in under-resourced urban schools where school day teachers may lack the time, resources, and skills to incorporate hands-on science into the day. A high school student shared an experience where she was able to speak with patients who suffered from colon cancer in the spring 2016 Pipeline Program to deepen her learning beyond books, "We know the symptoms and effects [of colon cancer] ... it's different to read about it, but to actually talk to that person face to face is something." In addition, a student in 2011 shared how a lesson required her to critically engage with problem-solving:

> Basically [the lesson was] like a case and they [the Penn students] would do the skit for the case. In our groups, we would think about what was wrong with the person. We were like doctors trying to figure out what was wrong with the person. And we learn what part of the brain affects what and why ... that was real fun to learn!

After school engagement thus serves as an ideal opportunity to enrich school day learning with culturally relevant science inquiry and provide students with varied learning environments that challenge their intellect and develop communication skills. One study revealed that Latinx students who struggled academically experienced a higher level of academic achievement in math than their higher-achieving peers when engaged in a problem-based STEM learning environment, in part due to new opportunities for student-driven intellectual exploration.[25]

The keys to offering quality care to diverse populations are a) strong communication skills, b) the development of trust between care providers and recipients and c) cultivation of authentic relationships with individuals, thereby disrupting hierarchical forms of leadership and embracing the vernacular,[26] challenging stereotypes of otherness and homogeneous conceptions of cultural groups.[27] With this, increased racial diversity among students in medical education can enhance the educational experience and learning for all students.[28] Health literacy is an equally important asset whether an individual is the caregiver or patient. The ability for an individual to share knowledge with others in a way that can be understood is critical to ensuring optimal health and well-being. Should high school students continue pursuing health-based or STEM careers, the ability to communicate is vital to future success and ensuring optimal personal and community health.[29] In addition, health literacy and the need for clear communication between health care providers and patients have been found to be a stronger predictor of health outcomes than age, income, educational level, or even race.[30]

The role of Penn students serving as science mentors helps to generate a learning context for mutual benefit and development of these skills. In the case of the Pipeline Program, mentors teach in small groups, where students can also teach and learn from their peers to enhance communication skills and increase understanding of the material.[31] A study conducted at a polytechnic school in Singapore examined the ways in which students across disciplines responded to problem-based learning facilitators. They found that the most important aspect to learning was a facilitator's level of "social congruence," or ability to informally and empathically communicate with students and foster an environment of open dialogue.[32] Youth shared in focus groups that they view the Penn students more as mentors than as traditional teachers. This created learning environment fosters a sense of community, which has been found to support learning and prevent unproductive behaviors, such as procrastination.[33] In addition, students are able to visualize themselves working in the mentors' disciplines, as one student shared about the spring 2016 Pipeline Program:

> My special moment [in the program] was actually meeting a real live cardiologist. I was like literally brought to tears, 'cause when I saw him, the feeling I had inside gave me motivation like I could really do this ... like become a cardiologist doing a lot of stuff with the heart. Seeing him and him telling me that [I could.]

These communities of practice serve as an intellectual and social space that directly shape student aspirations and identity while also increasing health literacy. Findings from the fall 2016 high school student survey reveal that having "a mentor who works in a STEM or health field" positively correlated to knowing about the educational pathway to become a doctor ($p < .001$); health issues that impact his/her community ($p < .001$); and confidence in being able to discuss health issues that could affect my family and friends ($p < .01$). Furthermore, the ability to listen, ask critical questions, understand key ideas and perspectives presented by others, clearly communicate ideas, and generate shared understanding while communicating in a variety of forms and contexts also enriches one's overall level of science literacy.[34] The Pipeline Program addresses this need through multi-generational engagement with Penn students during lessons and more specifically during the final presentations in the Spring program, proving valuable experiences for developing communication skills and reinforcing newly acquired content knowledge. One student in the spring 2015 Pipeline Program shared about having to give a formal final presentation to an audience, "it's alright because it's not a lot of people and I don't think everyone is used to having a lot of people to talk in front of."

Deepening Persistence and Engagement with STEM and Health-Based Pathways

While deeper engagement and STEM learning in high school are vital to students' success, we must also grapple with the reality that disparities in professional spaces persist due to structural barriers that block student success. The Pipeline Program and our participating high schools exist within this context, which requires us to continuously reflect on how to deepen collaboration and reflect on institutional changes necessary to achieve our goals. One study[35] tracked 33 ethnically and economically diverse high school students to explore why students who were once very interested in science, engineering, or medicine changed course or persisted in their pursuits throughout high school to develop their identity as a scientist. They found that students often perceived school-day science as hard and thus discouraging; however, exposure to science advocates at school or home and meaningful opportunities to work with science professionals improved positive attitudes and desires to continue pursuing their path in science, engineering, or health-based college or career fields.

The communities of practice, or *microclimates* students experienced at home, school, and outside of school had a direct influence on student trajectories. Notably, students who had more than one supportive space where they were enthusiastically invited to learn about science, value scientific ways of knowing, and/or pursue a degree or career in the sciences were more likely to continue to engage in their learning and pursuit of a STEM path. As students in the Spring of 2011,

2015 and 2016 Pipeline shared, the program was engaging and helped to connect students with valuable opportunities and mentorship: "I joined because I wanted to get to know the way around medicine, learning about more diseases ... I want to be a pediatrician;" and "I'm interested in psychology so thought learning more about the brain and how it works and functions would be good for that."

The majority of Pipeline Program students had also been engaged in other Netter Center programs at some point by their senior year, and findings reveal a positive correlation between increased engagement with different programs and an interest in science, inclination toward a health career, desire to continue education after high school, knowledge about health issues that affect family and friends, and confidence in being able to discuss those issues with family and friends ($p < .001$, all items). This finding is supported when comparing younger high school students who had only participated in the Pipeline Program in the Fall of 2016 to their peers who had also been in other Netter Center programs. Students who had been in multiple programs reporting more positively that "I am sure of myself when I do science" ($p < .05$). We posit that engaging in multiple extracurricular activities helps Pipeline students develop important skills, thus strengthening the impacts of the Pipeline Program to engage students with health issues and science application across programs.

In addition, based on our analysis, Pipeline students had significantly higher college matriculation rates than their peers who did not attend programs ($p < .01$). Half of the students from both high schools who enrolled in college during the Fall 2016 term had participated in the Pipeline Program, with 23% of those students pursuing either a science or health-based field. More importantly, however, Pipeline students demonstrated a significantly higher college term completion rate (92%) compared with their peers (86%), meaning that many students withdraw prior to successfully completing a semester ($p < .05$).[36] This is a promising finding as many of our West Philadelphia youth enroll in college, yet withdraw prior to completion of the semester.

Moreover, for the Classes of 2012 and 2014, controlling for enrollment in all other Netter Center programs, total enrollment in the Spring Pipeline Program was positively correlated with total terms of college enrollment, indicating that the program encouraged students to pursue post-secondary education. For the Class of 2012, one additional term enrolled in the Spring Pipeline Program, on average, increased college enrollment by 1.26 terms ($p < 0.05$). For the Class of 2014, one additional term enrolled in the Spring Pipeline Program, on average, increased college enrollment by 1.83 terms ($p < 0.05$).[37]

Other Netter Center programs that were positively correlated with total terms of college enrollment include the Extended Learning Afterschool and Summer

(ELAS) and College Access and Career Readiness (CACR) programs. For the Class of 2014, controlling for enrollment in all other programs, one additional term enrolled in the ELAS program, on average, increased college enrollment by 0.44 terms ($p < 0.01$). For the Class of 2015, one additional term enrolled in the CACR program, on average, increased college enrollment by 0.42 terms ($p < 0.001$). Most notably, however, the coefficients for the Spring Pipeline Program are three times larger than both ELAS and CACR.[38] This finding suggests that the Spring Pipeline Program attracts students who are motivated to pursue postsecondary education better than the other Netter Center programs. Alternatively, the spring Pipeline program may effectively influence students to enroll in college due to the low student-to-instructor ratio and activities that take place on Penn's campus.

In order to promote a broader applicant pool in health care-related fields, increase health literacy, and address the need to provide all patients with culturally-competent care, there needs to be a shift in representation of students from underrepresented communities entering and succeeding in post-secondary education. In a metrics-driven education system we, as well as our partners, are often faced with competing pressures to prove the merits of the Pipeline Program on academic gains over engaging in truly collaborative, transformative, and transgressive teaching and learning practices that enhance quality of life and social reform. Our next steps are thus to deepen our internal evaluation with a focus on implementation, student retention, more broad-based youth-led community outreach, and development of pedagogical strategies that continue to challenge the power balances too often found latent within medical education. In addition, we will continue to identify ways to expand and incorporate feedback from all participants to ensure that the program evolves and develops as an authentically collaborative and inclusive partnership, which has often proved to be a challenging endeavor.

Conclusion

In sum, barriers to culturally competent health care include: systems of care poorly designed to meet the needs of nondominant patient populations; poor communication between providers and patients of different racial, ethnic, or linguistic backgrounds; the need for a more inclusive and diverse medical education and emphasis on health literacy that begins at an early age; and lack of diversity in health care leadership and workforce.[39] An approach to help reach this ultimate goal, and broaden participation in STEM fields, is the development of scalable partnerships that focus on improving mathematics and science education and awareness for underrepresented students. In order to achieve this outcome, a need exists for more STEM and health mentors with both strong social congruence and content expertise to challenge youth intellect and development through inquiry-based, hands on science and community-health education. Programs like Pipeline can support youth engagement in their own learning and strengthen students' abilities to apply new knowledge as they address social health problems within local communities and in their homes. By learning about issues of importance, and working with Penn student mentors, students are provided opportunities to imagine pursuing careers where they can solve those issues.[40] Our hope is that findings shared from the Pipeline Program speak to the merits and potential of such a program and shed light onto our lessons learned regarding implementation to support successful replication efforts. Institutions of higher education have a critical role to play in helping prepare the next generation of students for college education, deepen non-dominant youth engagement in STEM and health-based educational and professional pathways, and address social and racial disparities in the real world through a replicable university-school partnership model.

Discussion Questions

- What kinds of systemic barriers exist for marginalized groups to access health care in your community?
- In what ways can we see issues of power shaping students' access to STEM pipelines?
- If we are able to better diversify the STEM pipeline, how can we correspondingly diversify health care leadership?

Notes

1. Psalm 90:10 (KJV, public domain)
2. A few days before he agreed to go to the doctor, he asked me—a doctoral student—what I thought about the tumor. Three days before he died, he asked me if he was going to be ok.
3. See Benjamin, 2017.
4. Cooper & Powe, 2004.
5. Berkman, Sheridan, Donahue, Halpern & Crotty, 2011.
6. American Association of Medical Colleges, 2012; AAMC 2014; AAMC 2016.
7. Engberg & Wolniak, 2010.
8. Neuhauser & Cook, 2016.
9. Young, Young & Paufler, 2017.
10. Netter Center for Community Partnerships, 2014. [Participant statistics]. Unpublished raw data.

11. StudentTracker provides attendance and completion data from more than 3,300 institutions representing over 92% of national postsecondary enrollment. Colleges that do not report attendance and completion data to the Clearinghouse are not included.
12. U.S. Census Bureau, 2015.
13. School District of Philadelphia, 2017.
14. Philadelphia Public School Notebook, 2015.
15. Roderick, Nagaoka & Coca, 2009; Koven, 2009; Cabrera, 2010; Plank & Jordan, 2001.
16. Roderick, Nagaoka & Coca, 2009.
17. Harvill, Maynard, Nguyen, Robertson-Kraft & Tagnatta, 2012.
18. Benson, Harkavy & Puckett, 2007.
19. Sturm, Eatman, Saltmarsh & Bush, 2011.
20. Brickhouse, 2001.
21. National Research Council, 2011; Seiler, 2001.
22. Basu & Calabrese Barton, 2007.
23. Kadlec, Friedman, & Ott, 2007.
24. Ullucci & Howard, 2015.
25. Han, Capraro & Capraro, 2014.
26. Weiner, 2003.
27. Gorski, 2016.
28. Whilta et al., 2003.
29. Gaglio, 2016.
30. American Medical Association, 1999; Kripalani & Weiss, 2006.
31. National Research Council, 2003; Lin & Hsi, 2000.
32. Yew & Yong, 2014, p. 796.
33. Twenge, Catanese & Baumeister, 2002.
34. Chung, Yoo, Kim, Lee & Zeidler 2014.
35. Aschbacher, Li, & Roth, 2010.
36. Netter Center for Community Partnerships, 2014. [Participant statistics]. Unpublished raw data.
37. *Ibid.*
38. *Ibid.*
39. Betancourt, Green & Carrillo, 2002.
40. Holdren, et al., 2010.

References

American Association of Medical Colleges. (2016). *Diversity in medical education: Facts and figures 2016*. Retrieved from http://www.aamcdiversityfactsandfigures2016.org/

American Association of Medical Colleges. (2014). *Diversity in medical education: Facts and figures 2014*. Retrieved from http://aamcdiversityfactsandfigures.org/

American Association of Medical Colleges. (2012). *Diversity in medical education: Facts and figures 2012*. Retrieved from https://members.aamc.org/eweb/upload/Diversity%20in%20Medical%20Education%20Facts%20and%20Figures%202012.pdf

American Medical Association. (1999). Health Literacy: Report of the Council on Scientific Affairs. *Journal of the American Medical Association, 281*(6), 552–557.

Aschbacher, P. R., Li, E., & Roth, E. J. (2010). Is science me? High school students' identities, participation and aspirations in science, engineering, and medicine. *Journal of Research in Science Teaching, 47*(5), 564–582.

Basu, S. J., & Calabrese Barton, A. (2007). Developing a sustained interest in science among urban minority youth. *Journal of Research in Science Teaching, 44*(3), 466–489.

Benjamin, R. (2017). Cultura obscura: Race, power, and "culture talk" in the health sciences. *American Journal of Law & Medicine, 43*(2–3), 225–238.

Benson, L., Harkavy, I. R., & Puckett, J. L. (2007). *Dewey's dream: Universities and democracies in an age of education reform: Civil society, public schools, and democratic citizenship*. Philadelphia, PA: Temple University Press.

Berkman, N. D, Sheridan, S. L, Donahue, K. E., Halpern, D. J., & Crotty, K. (2011). Low health literacy and health outcomes: An updated systematic review. *Annals of Internal Medicine, 155*(2), 97–107.

Betancourt, J. R., Green, A. R., & Carrillo, J. E. (2002, October). *Cultural competence in health care: emerging frameworks and practical approaches*. Retrieved from http://www.commonwealthfund.org/~/media/files/publications/fund-report/2002/oct/cultural-competence-in-health-care--emerging-frameworks-and-practical-approaches/betancourt_culturalcompetence_576-pdf.pdf

Brickhouse, N. W. (2001). Embodying science: A feminist perspective on learning. *Journal of research in science teaching, 38*(3), 282–295.

Cabrera, K. E. (2010). *Sustaining success toward closing the achievement gap: A case study of one urban high school*. (Doctoral dissertation). Retrieved from Proquest Dissertations & Theses A & I (Order No. 3403531).

Chung, Y., Yoo, J., Kim, S. W., Lee, H., & Zeidler, D. L. (2016). Enhancing students' communication skills in the science classroom through socioscientific issues. *International Journal of Science & Mathematics Education, 14*(1), 1–27.

Cooper, L. A. & Powe, N. R. (2004, July). *Disparities in patient experiences, health care processes and outcomes: The role of patient provider racial, ethnic, and language concordance*. Retrieved from http://www.commonwealthfund.org/~/media/files/publications/fund-report/2004/jul/disparities-in-patient-experiences--health-care-processes--and-outcomes--the-role-of-patient-provide/cooper_disparities_in_patient_experiences_753-pdf.pdf

Engberg, M. E., & Wolniak, G. C. (2010). Examining the effects of high school contexts on postsecondary enrollment. *Research in Higher Education, 51*(2), 132–153.

Gaglio, B. (2016). Health literacy—An important element in patient-centered outcomes research. *Journal of Health Communication, 21*(Supp. 2), 1–3.

Gorski, P. (2016). Rethinking the role of "culture" in educational equity: From cultural competence to equity literacy. *Multicultural Perspectives, 18*(4), 221–226.

Han, S., Capraro, R., & Capraro, M. M. (2014). How science, technology, engineering, and mathematics (STEM) project-based learning (PBL) affects high, middle, and low achievers differently: The impact of student factors on achievement. *International Journal of Science and Mathematics Education, 13*(5), 1089–1113.

Harvill, E. L., Maynard, R. A., Nguyen, H. T., Robertson-Kraft, C., & Tognatta, N. (2012). *Effects of college access programs on college readiness and enrollment: A meta-analysis.* Cambridge, MA: Abt Associates. (ERIC Document Reproduction Service No. ED530404)

Hill, D. H. (2008). School strategies and the "college-linking" process: Reconsidering the effects of high schools on college enrollment. *Sociology of education, 81*(1), 53–76.

Holdren, J. P., Lander, E. S., & Varmus, H. (2010). *Prepare and inspire: K–12 education in science, technology, engineering, and math (STEM) for America's future. Executive Report.* Washington, DC: President's Council of Advisors on Science and Technology. Retrieved from https://nsf.gov/attachments/117803/public/2a--Prepare_and_Inspire--PCAST.pdf

Kadlec, A., Friedman, W., & Ott, A. (2007). *Important, but not for me.* Brooklyn, NY: Public Agenda. (ERIC Document Reproduction Service No. ED498649) Retrieved from https://files.eric.ed.gov/fulltext/ED498649.pdf

Koven, K. A. (2009). *Establishing college preparatory conditions and a college-going culture in California charter high schools.* (Doctoral dissertation). Retrieved from Proquest Dissertations & Theses A & I (Order No. 3405599).

Kripalani, S., & Weiss, B. D. (2006). Teaching about health literacy and clear communication. *Journal of General Internal Medicine, 21*(8), 888–890.

Linn, M. C., & Hsi, S. (2000). *Computers, teachers, peers: Science learning partners.* Mahwah, NJ: Erlbaum: Routledge.

National Research Council. (2003). *Engaging schools: Fostering high school students' motivation to learn.* Washington, DC: The National Academies Press.

National Research Council (2011). *Successful K–12 STEM education: Identifying effective approaches in STEM.* Washington, DC: National Research Council of the National Academies.

Netter Center for Community Partnerships (2014). [Participant statistics]. Unpublished raw data.

Neuhauser, A. & Cook, L. (2016, May 17). *2016 U.S. News/Raytheon STEM index shows uptick in hiring, education.* Retrieved from https://www.usnews.com/news/articles/2016-05-17/the-new-stem-index-2016

Philadelphia Public School Notebook. (2015, May 18). *Philadelphia's graduation and college-going rates, school by school.* Retrieved from http://thenotebook.org/latest0/2015/05/18/philadelphia-s-graduation-and-college-going-rates-school-by-school

Plank, S. B., & Jordan, W. J. (2001). Effects of information, guidance, and actions on postsecondary destinations: A study of talent loss. *American Educational Research Journal, 38*(4), 947–979.

Roderick, M., Nagaoka, J., & Coca, V. (2009). College readiness for all: The challenge for urban high schools. *The Future of Children, 19*(1), 185–210.

School District of Philadelphia. (2017). *School Profiles.* Retrieved from https://dashboards.philasd.org/extensions/philadelphia/index.html

Seiler, G. (2001). Reversing the "standard" direction: Science emerging from the lives of African American students. *Journal of Research in Science Teaching, 38*(9), 1000–1014.

Sturm, S., Eatman, T., Saltmarsh, J., & Bush, A. (2011). Full participation: Building the architecture for diversity and public engagement in higher education. *White Paper, Columbia University Law School, Center for Institutional and Social Change.* Retrieved from http://imaginingamerica. org/wp-content/uploads/2011/10/Catalyst-Paper.pdf

Twenge, J. M., Catanese, K. R., & Baumeister, R. F. (2002). Social exclusion causes self-defeating behavior. *Journal of Personality and Social Psychology, 83*(3), 606–615.

Ullucci, K., & Howard, T. (2015). Pathologizing the poor: Implications for preparing teachers to work in high-poverty schools. *Urban Education, 50*(2), 170–193.

U.S. Census Bureau. (2015). *Social explorer—ACS 2015 (5-year estimates) select characteristics for four West Philadelphia 5-digit zip codes (ZCTA5)*. Retrieved from http://www.socialexplorer.com/tables/ACS2015_5yr/R11326398?Re portId=R11326398

Weiner, E. J. (2003). Secretary Paulo Freire and the democratization of power: Toward a theory of transformative leadership. *Educational Philosophy and theory, 35*(1), 89–106.

Whilta, D. K., Orfield, G., Silen, W., Teperow, C., Howerd, C., & Reede, J. (2003). Education benefits of diversity in medical school: a survey of students. *Academic Medicine, 78*(5), 460–466.

Yew, E. H. J., & Yong, J. J. Y. (2014). Student perceptions of facilitators' social congruence, use of expertise and cognitive congruence in problem-based learning. *Instructional Science, 42*(5), 795–815.

Young, J. L., Young, J. R., & Paufler, N. A. (2017). Out of School and into STEM: Supporting Girls of Color through Culturally Relevant Enrichment. *Journal of Interdisciplinary Teacher Leadership, 1*(2).

CHAPTER NINE

Institutional Capacity Building for STEM Teacher Education at an Urban Commuter University

JANELLE M. JOHNSON,[1] ROLAND SCHENDEL,[2]
ELIZABETH McCLELLAN RIBBLE,[3] AND HSIU-PING LIU[4]

This research was supported by NSF grant #DUE-1540805

Abstract

This chapter is an outcome of research conducted under an NSF Robert Noyce Capacity Building grant during 2015–2016. With MSU Denver's 2014 Equity Scorecard Report[1] as its framework, the study's overarching research question examined what various program stakeholders viewed as key elements of a scholarship

---

1 Janelle M. Johnson, Assistant Professor of Secondary Education
  Metropolitan State University of Denver
  jjohn428@msudenver.edu
2 Roland Schendel, Assistant Professor of Literacy, Elementary Education & Literacy
  Metropolitan State University of Denver
  rschende@msudenver.edu
3 Elizabeth McClellan Ribble, Associate Professor of Mathematics
  Metropolitan State University of Denver
  emcclel3@msudenver.edu
4 Hsiu-Ping Liu, Professor of Biology & Director of Center for Advanced STEM Education
  Metropolitan State University of Denver
  hliu1@msudenver.edu

program designed to serve its urban commuter population. Data were collected with stakeholder groups including university faculty, student support services providers, STEM students, Community College of Denver faculty, community-based organizations, established Noyce programs, school/district partners, students at partner secondary schools, and School of Education administrators. Each group provided detailed recommendations for the program design. This chapter specifically focuses on the voices of the university students in STEM majors, some of whom are pursuing teacher licensure.

Editorial reflections

The first time I, Joy, saw the Rocky Mountains up close was as a fellow for a AAAS Media program. Working at a local NPR station, I wrote and produced science news stories for the morning show. In 1992, the notion of "STEM" had not quite been birthed into public lexicon yet even though "Do the Right Thing" had. I sported large braids and kente-printed fabric shirts. I produced stories about controversial stem cell research that is now well within the limits of public understanding. I also wrote a story about new technology (the polymerase chain reaction) and the human genome project. The conservative community was not quite ready for me and let me know that they were instead quite curious about me—I became the "other" among them but I didn't know it at the time. Eventually the novelty of otherness wore off; my job became more difficult to do and my commitment to the program waned. One positive thing I took from Denver was an appreciation for the conversation process, the dialogue. Curiosities about me prompted people at the station to ask questions ... often. After that summer, I never heard from them again, but, I had no expectations. It was a six-week program.

Fast-forward twenty-five years and the human genome project has gained relevance within the region in the form of ancestry testing by the genetic data mining industry. Twenty-five years after my first experience in the region, Denver, Colorado presents as one of the most progressive cities in the West. Marketed as a young, active city at the base of the Colorado Rockies, Denver is diverse. It feels cosmopolitan and ready (now) to move from diversity iteration into equity operation—transgression—the execution of greater, more meaningful, collaboration rather than mere admiration of "other."

STEM equity seems to be a pretty hot topic in general at the moment. The National Science Foundation has a huge emphasis on broadening participation in STEM. Industries are studying demographic projections and panicking about who their future workforce will be. They announce major "diversity" initiatives and fund varying outreach programs to target communities of color. In Colorado, this is evidenced by the presence of organizations like the STEM Champions whose stated mission is to diversify STEM;

designations are made by funding levels rather than any knowledge of inclusive practices. I, Janelle, could not access the group's meetings though I tried. Other organizations, recognizing the need for an educator with experience in equity, have invited me to do work as a consultant. I always feel a combination of honor to be asked, and enthusiasm that an organization is willing to tackle this huge challenge. There have, however, been some hard lessons to come out of this work. Any respectable equity researcher will tell you there is no recipe for inclusive pedagogy. While there are certainly practices known to be effective, practitioner transformation only comes over time, with sustained and often uncomfortable work. This same principle needs to be considered with organizations that are seeking greater diversity and inclusion.

Equity is not something that can be overlaid on a problematic foundation; I was reminded of this during some consulting work I did recently. I'm embarrassed to say that the highly contentious interactions that arose caught me off guard. Despite the messaging from so many academic and professional sources that tell me I need to claim expertise, as a White practitioner of equity, humility is a far more useful approach. I am not going to "pop in and fix" some organization with a day or a week of "training." I know that intellectually, of course, but I had let myself become too comfortable. After that problematic experience I was anxious to return home, not just out of emotional exhaustion, but because I realized that the work I was returning to was designed from the bottom up on a foundation of equity and inclusion. That work has its challenges too, of course, but the long-term nature of the project means that we can learn from tensions that arise, and account for them in our structures, rather than just trying to "manage" a crisis as happens during the short-term. Equitable STEM can only be the product of authentic investment by all its stakeholders. The capacity building project you will read about in the upcoming chapter tells the story of that slow, intentional work. Reflecting on these very different scenarios helped us to re-realize the importance of addressing racism at a structural level in our own work. No quick fix exists.

Introduction

In an era when the importance of science, technology, engineering, and math learning is broadly recognized, many students are underprepared and/or unmotivated to pursue STEM studies and careers. Other students who demonstrate interest or ability are pushed out at multiple points along the cradle to career trajectory described as a "pipeline." The outcomes of both these scenarios are typically framed as an "achievement gap," there are multiple historic systemic contributors to these troubling outcomes and leaky pipelines. Limited exposure to STEM learning is confounded by many factors associated with poverty: limited access to supplemental educational programs; lower quality schooling; poor teacher expectations that reinforce stereotype threat; reduced parental education and involvement; cultural and language differences from what is being assessed in schools; peer influences that can constrain aspirations; and lack of understanding of pathways to and through higher education. Students who come from communities underrepresented in STEM fields face what is better described as an opportunity gap.

This chapter describes work that was designed to illuminate pathways into STEM and STEM teaching with the opportunity gap in mind. It begins with a description of the context where the work and the research took place—an urban commuter campus called Metropolitan State University of Denver. The early section of the chapter also reviews some of the research literature that was utilized to write a successful grant proposal and build a foundation for future grant-funded programming. Next is a brief description of the research methods utilized in the study, followed by a thematic outline of the research findings. Though the research collected data from many of the university's stakeholders, this chapter focuses on the voices of the undergraduate students. Some discussion is embedded with the findings, and is continued in the final section along with recommendations. The idea of offering recommendations is not to be prescriptive, but rather to offer some lessons learned from our own work that may serve as promising practices to try out in other contexts.

Research Context and Review of the Literature

Metropolitan State University of Denver (MSU Denver) was established in 1965 as Metropolitan State College with the mission of providing skilled workers to the Metro Denver area, and its policy of open admission continues—93.5% of students enrolling at MSU Denver today are local residents. In 1968, funding created a campus shared with University of Colorado-Denver and Community College of Denver, an action that displaced primarily people of color in the Auraria neighborhood.[2] This legacy of simultaneously serving and underserving Denver communities of color is a thread that weaves through the contemporary context. While noted for its relative diversity, there are problematic inequities in the university's rates of persistence and graduation. MSU Denver has long been recognized locally for providing an affordable, high quality education in Colorado. and passed the mark to be designated as a Hispanic Serving Institution (HSI) in fall 2017; forty percent of students, 33% of full-time staff, and 25% of full-time faculty are people of color. MSU Denver is socioeconomically diverse as well. Thirty-four percent of students are low income (family income less than 150% of federal poverty level), and 32% are the first generation in their families to attend college. The vast majority of students work part or full time jobs while attending school to mitigate loan debt and cover living expenses and family responsibilities. Approximately 80% of alumni remain in metro Denver.

Traditional metrics based on four year residential colleges would fail to capture the multiple pathways MSU Denver students take into and through the university. Among undergraduate students, nearly 40% attend school part time, and the average student takes five years to graduate. Many students take far longer. Our students have a higher average age than students in other Colorado universities since many of them come back to school after another career, time in the military, or attending to family obligations. Because of MSU Denver's relative affordability compared to other universities in Colorado, the student population is more diverse—a similar phenomenon to what we observe in community colleges. However, finances remain a huge roadblock for most of our students. It was this fact that motivated our team to apply for National Science Foundation (NSF) Noyce funding for a Scholarships and Stipends grant—60% of the grant money would go directly to students.

During the past decade, MSU Denver became part of a nationally supported initiative called *Equity in Excellence* in partnership with University of Southern California's Center for Urban Education[3] and the Colorado Department of Higher Education. Its aim was to analyze campus-level data to make recommendations for addressing disparities in academic outcomes for underrepresented groups, and shed light on at least some of the university's own shortcomings. The university began to take actions in response to recognition of the economic and social challenges its students face, pioneering a policy of offering discounted tuition for undocumented students, striving for increases in graduation rates for African-American and Latinx students, and growing student-support programs and staff training and development. As our team began to craft a grant proposal for National Science Foundation funding, the leadership team decided to utilize the university's *Equity in Excellence* report[4] as a framework. The document represented a lot of collaborative work at the university that resonated with the team's effort for greater integration across and between programs and departments, while acknowledging the structural barriers that may have shaped students' access to high quality education during their PreK-12 education.[5]

Such systemic barriers have always served to marginalize certain groups, but in our work with high need schools and college students who are the product of those schools, the underpreparation of certain communities of students is notable. In Colorado, "essential skills" including critical thinking and reasoning, information literacy, collaboration, self-direction, and invention, have been described as preparing students for future careers that have not even been conceived yet. In Colorado, the statewide roadmap vision of STEM education "focused on building critical and creative thinking and analysis skills by addressing how students view and experience the world around them. Strong STEM teaching and learning opportunities rest on inquiry-, technology-, and project-based learning activities and lessons that are tied to the real world; a diverse, interdisciplinary curriculum where activities in one class complement those in other classes."[6] While certainly a laudable goal, access to this type of education is extremely uneven.

The National Science Foundation consistently reiterates its objective to broaden participation in STEM fields, expand the scientific literacy of all citizens, increase participation of underrepresented groups and underrepresented institution in STEM, and build a more diverse workforce.[7] Though one way NSF strives to support diversity is in the types of institutions it awards grants to, the infrastructures of large, research universities are generally better suited to receiving and running those grants than smaller public teaching institutions such as ours. We applied for a capacity building grant *because we needed to build institutional capacity* … we would have been completely underprepared to dive immediately into running the scholarship program. Despite our structural limitations, our university seems to provide some insight into future educational trends. When viewed against a context of skyrocketing college tuition nationally, a student loan crisis that continues to intensify (especially for teachers),[8] and demographic projections of a U.S. population continuing to become more diverse (racially, ethnically, *and* linguistically), funding research to learn about students' needs at an urban commuter campus does make sense.

It seemed clear from the comments that NSF reviewers appreciated the diverse demographics of our institution coupled with the equity-based nature of our proposal, and we were awarded one of the one-year capacity building grants. Our stated goals in the proposal included cultivating strong relationships with high-needs school districts; we took on this task by conducting district math and science teacher gap analyses, working to embed partner districts' STEM teaching needs into our teacher preparation coursework, and establishing a range of high quality clinical placement sites. Our second goal was to align our math and science licensure programs with our Noyce program design. Our approach was to modify the design of clinical structures and build more integrated and interdisciplinary STEM learning experiences for future teachers, including a year-long teacher residency and monthly STEM seminars. Our third goal was to develop a teacher licensure track for physics majors. The physics department worked to integrate more engineering design into their coursework, update their laboratories, and in conjunction with math faculty, establish a Learning Assistant

Program[9] in STEM courses. Goal four was to research and develop recruitment and retention strategies first for STEM majors, and then for STEM majors to consider becoming math or science teachers. We felt that if we reached these four goals, it would indicate that we had the institutional capacity to host a Noyce Scholarships & Stipends program—write the proposal, create needed administrative support structures, consult regularly with a broad advisory network, and secure outside evaluation.

Research Methods

The U-STEM research methodology represents a mixed methods case study[10] based on MSU Denver's different stakeholder groups and their varied perspectives. Qualitative data collection methods included semi-structured focus groups with specific questions and open-ended questions with follow-up questions based on participant responses. Quantitative data collection focused on participant questionnaires, demographic data, student trajectories and graduation rates over a five-year period. Students participating in the survey were largely recruited through the Colorado Wyoming Alliance for Minority Participation (CO-WY AMP) program since it serves as the foundation for the Noyce program. As you can see in the graphs below, respondents to the survey reflect a higher proportion of non-White students than the university overall in 2015.

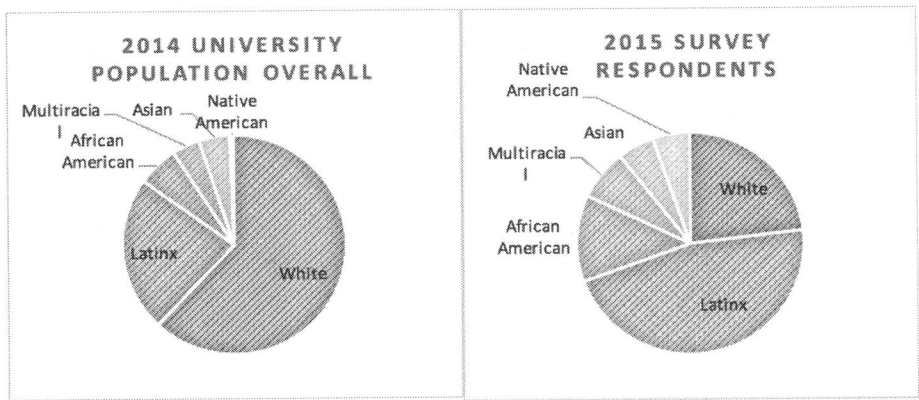

Figure 9.1: Overall university population at Metropolitan State University of Denver compared to survey respondents.
Source: Johnson, J., Liu, H.P., Bernhardt, P.B., Evans, B., Koester, M., & Loats, J. (2016). [Student survey]. Unpublished raw data.

Data were analyzed by both statistical analysis and thematic analysis and coding of research participants' narratives. Similar to the constant comparative method,[11] portraiture methodology was used to understand participants' insights and identify key ideas.[12] As underlying conceptual patterns and ideas were revealed, themes consequently began to take shape. Exploration of participant experiences and perceptions included an open coding process[13] designed to bring the data to life, allowing the analyst to examine what and how participant views were shared. Collected data were sifted to tease out patterns and themes for organization. Then the narrative that contributed to the final qualitative analysis was constructed.[14]

Findings

The findings are organized into the themes that emerged from the data analysis that reveal the need for multiple changes at the institutional level: STEM courses themselves acting as gatekeepers; student uncertainty about STEM; influences that shape STEM pathways; and supports that students named as facilitating their learning in STEM.

Gatekeepers and Leaky Pipelines

The phenomenon of STEM courses and STEM professors acting as gatekeepers is well documented in the research. Students most negatively impacted are first generation students and students of color, with a potential domino effect of negative implications on efforts by schools of education to diversify their pool of STEM teacher candidates. During the capacity building research, students expressed the gatekeeping nature of STEM courses coupled with a need for greater support. One MSU Denver Physics student explained, "One of my lab partners was taking physics, and her instructor was more concerned with people being able to get symbols right than actually understanding the problem." The student went on to question the value of the experience and the instructor's purpose. "You can learn the letters, you'll eventually learn the letters. But, can you do the math? Can you actually learn the process? I think approach is very important." At MSU Denver, the Center for Advanced STEM Education (CASE) and CO-WY AMP serve to mitigate gatekeeping effects; these supports are described in the *Facilitation of Learning* section below.

Finances and financial constraints were found to be another important factor for students' STEM learning. Due to outside work commitments, students reported a limited ability to engage in academic support structures and community-building activities, both of which are factors known[15] to be crucial for first

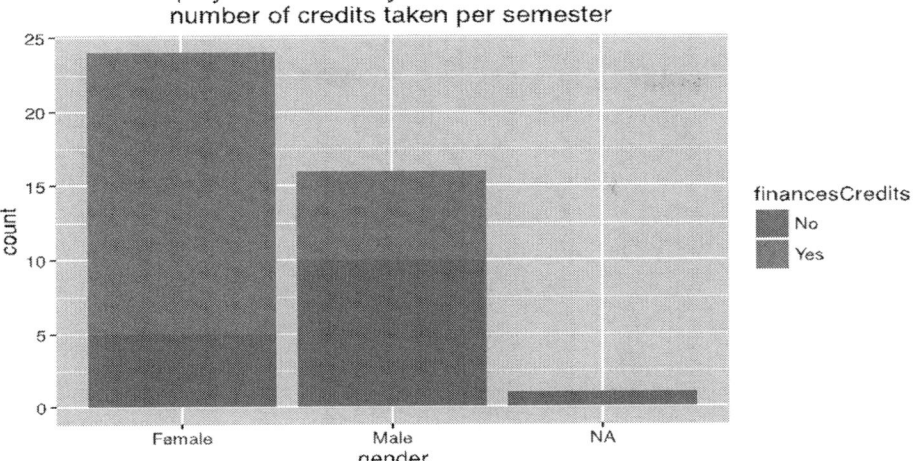

Figure 9.2: Survey question asking how finances have played a role in students' education at MSU Denver.
Source: Johnson, J., Liu, H.P., Bernhardt, P.B., Evans, B., Koester, M. & Loats, J. (2016). [Student survey]. Unpublished raw data.

generation student success.[16] It also impacted the number of credits students were able to take per semester; this was especially the case for women.

In terms of attracting STEM majors to consider becoming teachers, the pipeline is leaky for all students, and especially students of color. At MSU Denver, both math and science licensure students reflect a nearly 50–50 gender balance, but race/ethnicity is skewed White: students of color comprise only 25% of math and 19% of science licensure students, as compared to 29% for the School of Education, 30% for the math and 36% for the science majors, and 38% for the University overall. Over the past five years, the number of undergraduates completing their teacher licensure at MSU Denver averaged 6.8 per year for mathematics and 8.8 per year for science; students of color completing their teacher licensure in STEM fields represented an average of 1 per year for math and 1.6 per year for science.[17] Many STEM majors at MSU Denver reported an interest in teaching but were unsure of the procedures to do so. One student wrote, "Simply presenting it as an option is a HUGE start. Few consider it because it's not presented. Also, make it clear how to get licensed as a teacher." These conditions reveal a need for a shift in recruitment and retention of preservice teachers by the School of Education. Other STEM majors described the low pay and prestige of teaching as a disincentive to considering the profession.

Uncertainty about STEM as a Pathway

Developing greater understanding of what a STEM pathway has to offer was of key interest to participating students. The students who participated in the capacity building research explained the importance of early exposure and information about what STEM was, and why it may be a good fit for them. "I didn't start researching about STEM until my freshman year of college," explained one student. Others said, "You can't plan for something if you don't even know it exists," and "I wish I knew more information about STEM while I was in high school." One MSU Denver student suggested that the university offer an exploratory STEM course:

> One thing that I always thought would have been nice was a class I could have taken that was ... STEM. This is your introduction to STEM [class] itself, to let you test out all the different paths in STEM without having to actually devote towards any one of them at that point. You could come in and experience what engineering is like on a college level, this is what mathematics is like at the college level, without having to actually take math for non-majors, biology for non-majors ... it's a lot of classes! If there was one class that could give you that introduction, then it would be a lot easier to say, "I really liked that part of the class, and I would like to go that way, or I really liked this part of the class, and I could go that way." I really wish that there was something like that.

Clarifying what STEM is can open up more pathways into STEM and STEM teaching. A current student at MSU Denver in the math and science program revealed the support that a clear definition gives to the potential of a program,

> When I was a kid, STEM was just science and math. That was it. Now I'm working with kids and I view it as an avenue for our kids to really create and become innovators and inventors. There is an added sense of practicality and usability in STEM in that their creations and their innovations go out to the masses to be used to help society and put science into practice.

This connects with research that shows the importance of STEM relevance and potential for application that helps counter trends for underrepresented students, helping to create multiple pathways into STEM and STEM teaching.

Student Pathways into STEM and STEM Teaching

First generation students consistently express the need for additional assistance navigating into and through the institution. Beyond expressing the need for advising to be better, more frequent, and more consistent, MSU Denver student participants offered practical ideas based on other effective advising programs on campus.

> There are programs out there. Athletics are the most overt ... you almost need a support system for STEM folks ... you could have a case manager that helps you through ... TRiO[18] has it, military folks have it ... like you have your own STEM counselor that helps with your classes.

Students expressed a strong need for mentoring and information transparency, issues that have measurable implications for recruitment, retention, and graduation rates. There is also a need for smoother transitions between institutions—high schools, community colleges, and four year universities.

While academic advising is essential to effective navigation of the college learning process, peer support can add an additional level of support to students. Several student participants articulated the value of having a knowledgeable peer helping them. One student reflected, "Thankfully I have a friend who had already graduated from Metro so he helped me get through the process." Another student offered a suggestion for support, "I think it might help to actually talk to some people who are in the process of college entry. If you're not thinking STEM, why aren't you thinking STEM?" Student support and informed advisors together might serve to overcome the challenges associated with planning college courses by boosting student efficacy.

Based on the research literature on first generation college students, we expected to see a strong family influence on student decision making.[19] A current MSU Denver physics student described how he would share his potential participation in the Noyce U-STEM program, saying,

> I would convince my family by showing them something like this, the brochure. I would tell them, "Look, I know this isn't the path of an engineer, but understand that I still will be able to learn, critically think about becoming an engineer." I would have to look into the details of this before explaining to my parents. I would have to tell them like this, "Well, I looked into the details ... MSU Denver doesn't just provide an education degree. It's a mathematics degree with an education concentration."

Several students expressed uncertainty about what teacher licensure entailed, or what the procedures were. In terms of selecting a STEM major, however, students reported making that decision *without* family involvement; nearly 88% reported that a family member was *not* an influence.

Finances were found to be a severely limiting factor for students to consider a transition from a pure STEM major to secondary STEM teaching licensure. Research shows finances take on an especially important role for first generation college students.[20] Students were frank in expressing their hesitation to consider teaching. Responses included "Promise of comparative compensation to that of healthcare and engineering jobs. Incentives to become a teacher rather than those possible six-figure jobs;" "Fix the education system;" "Pay them more. I know it's not up to you but it's the problem;" and "Make teaching pay more and get rid of standardized testing."

Students also reported that they would have liked to have had conversations as middle and high school students with their teachers that would have supported their understanding of the opportunities for following STEM career paths or potentially becoming a teacher. They described how that would have been a way for their teachers to express not only high expectations for them, but it also would have showcased the qualities and achievements necessary for following such pathways. This would allow students to proactively prepare, mentally and academically, for the STEM career pathway. This finding emphasized the importance of working with classroom teachers to strengthen the pipeline for students to pursue STEM and STEM teaching paths. Specific to preparing STEM majors to become teachers, a physics student reflected on the value of using inclusive teaching strategies to meet the needs of diverse students.

> I think that's so great to receive training in culturally and linguistically diverse and relevant teaching methods, because a lot of students who don't speak English or are English learners start to think that English is most important, instead of their overall learning. It's about how to be involved in our English community. That's where math and science needs to be, where they [students] can critically think about that.

While MSU Denver does have multiple outreach programs in place in Denver area schools, because of its open admission policy, specific recruiting efforts are extremely limited. Our proposed scholarship program design created a network of near-peer mentors to connect students at the university, the community college, and secondary schools. Students responded positively to this idea: "I think this program would really help in doing that because different types of science majors and math majors can go into those high schools and tell them this is what we're going to be doing. This concept relates to this kind of career or this field of study." Another student declared,

> If I was a freshman and I saw this, I'd be like, "This is the program that I want to do. This is what I want to do." If you ask me why [I'm] going to do this, I would say: "One, I really enjoy math. Two, this is a really great and organized plan for someone who wants to go into math education."

Facilitation of Learning in STEM

Our research revealed a longstanding need to improve course sequences for math and science majors and math and science licensure candidates. More than 70% of STEM majors reported that the way their courses were sequenced were not helpful in progressing through their major. There were similar disconnects for licensure students—some of the math and science courses required for pre-service teachers were not the courses the home department recommended for content knowledge development. For example, the traditional course sequence for non-chemistry science licensure students was to take a principles of chemistry course that is actually geared toward nursing majors, followed by an introduction to organic/biochemistry course. Teachers in Colorado are licensed to teach "science" rather than in a specific discipline like biology or physics. Though a student has to major in one content area, they must be prepared to teach any science course. The perspectives of chemistry department faculty were gathered while conducting capacity building research. They felt strongly that the courses that would best prepare future science teachers are General Chemistry I and II. The capacity building process facilitated conversations and curricular changes between and across STEM departments and the School of Education, resulting in greater awareness of the need to purposefully streamline course sequences to better meet future math and science teachers' needs.

University STEM professors' use of more inclusive pedagogies, including utilization of Learning Assistants in many courses, also helped to facilitate student learning.[21] A student explained, "We had to do math learning designs in groups. It really helps because then we can achieve greater learning" through others. Another student explained, "Collaboration is key. When we tell each other our manipulations and how we got our answer and everything, it really shines a light. It helped me see that I can teach this really well to help others understand." Students underrepresented in STEM fields demonstrate much greater success in supportive cohort models;[22] the structure of the Learning Assistant program allows students to gain other perspectives and to practice "teaching" others. Exposing students to teaching and learning has the potential to open STEM teaching pathways. When asked what kinds of experiences could help draw STEM majors to teaching, one student offered,

> Tutoring courses/workshops would be a good start. Most people don't know that they like teaching, or don't feel good at it. Maybe a day where kids/teens come to the school and we get to present cool science things to them, or have an interactive day. I would do that!

Purposefully merging courses is also a possibility according to some participating MSU Denver students. One student spoke of a learning experience of a peer and how such an experience might benefit her. She explained,

> They actually created a biology-chemistry merge class recently. I wish I could have taken it! So they essentially took the first semester of both and merged them together so you could learn them both simultaneously, and you could see all those overlaps. You actually see why these things are important, together. It's hard to connect them if you take them separately.

Collecting insights from students about the potential value of merged courses such as this might serve to meet the learning needs of individuals, support their content area understanding, and strengthen their educational experiences, broadening access to STEM and STEM education pathways.

Discussion and Recommendations: Promising Practices

Our university may have offered a promising site for NSF funding probably because of its emerging HSI status and our proposal's equity-based framework. As pointed out by many authors, however, a context of d*iversity* does not imply *inclusion*.[23] Just as research that demonstrates that increasing the number of women or minorities will not close so-called achievement gaps,[24] a teacher preparation program housed in an institution of racially diverse students does not necessarily produce more teacher candidates of color. We do not claim to have any answers, but we have engaged in a process of critical institutional self-reflection, or praxis. This list of recommendations represents a synthesis of actions to take based on our capacity building research and connects back with some of the related research literature. While we incorporated each of these actions in our proposal for a five-year Noyce Scholarship grant that was successfully funded, we are still in year one of that funding. While we certainly *intend* to follow these recommendations ourselves, time and experience will offer further insights.

- Develop early recruitment strategies with secondary students and their families, and with first and second year students at community colleges and universities to increase information dissemination of STEM and STEM teaching pathways.[25]
- Offer a variety of co-curricular activities designed to promote students' sense of STEM identity and foster academic success. Labeled as "service" as part of receiving the scholarship funds, this includes participation as a Learning Assistant; working in the Summer Science Programs for middle and high school students doing hands on activities and research with MSU Denver faculty; tutoring; serving as a peer ambassador at a community college, high school, or middle school; volunteering with community-based organizations; or allowing the student to design their own service activities that they are passionate about.
- Offer professional development for faculty and staff through learning communities and targeted sessions. Members of the Noyce team ran a faculty and staff learning community on STEM equity, and the Equity

Coordinator is currently working with various STEM departments to offer regular workshops.
- Provide student participants with the following resources: monthly skill development meetings; study groups; free tutoring services; and ensure student accountability for program requirements. Of the students who responded to our capacity building survey, while only a small fraction of White students reported benefitting from CASE tutoring, about half the Latinx students described the tutoring as a benefit.
- Tailor student advising within each STEM major by connecting students with specific advisors who are prepared with information on programs and have a proven record of support of underrepresented student populations.
- Past advising gaffs and/or lapses in university enrollment over time may leave students with gaps or missing credits as they fall between the cracks of catalog years. Closely track student progress toward degree completion with available software tools. Keep in mind that such standardized institutional-level systems may reinforce deficit views of students, labeling them as "at risk" and conveying lowered expectations.
- Track students throughout their education through degree completion and beyond by providing guidance and targeted strategies for success. We created a position called a "Student Success Professional" to focus on doing this work, helping guide students successfully into, across, and out of the institution and into the workplace. The human element, especially based on relationships developed with students over time, helps students access institutional knowledge that can help them successfully navigate the institution.
- Build and/or participate in various programs proven to increase retention: internships;[26] undergraduate research projects and conferences;[27] the Learning Assistant program;[28] tutoring and mentoring programs;[29] serving as mentors for youth in summer science camps;[30] and working with faculty on research projects.[31]
- Fund and support students attaining professional organization memberships and attending and presenting at regional and national math and science conferences. This is an opportunity to be exposed to best teaching practices and the latest research, as well as network with teachers and administrators from local school districts. Conference presentation teams will strive to also include university and partner school faculty.
- Make sure that students work closely with an advisor in his or her science or mathematics content area and ensure regular meetings. Though this is officially required, we learned during capacity building research that it occurs irregularly. We have taken the extra step of highlighting which advisors are

both familiar with our program and better at responding to the needs of students who are first generation and/or from underrepresented communities.
- Tap into existing organizations and structures that are already working effectively with target populations for both student recruiting and internship sites. In our case these include community based organizations such as Colorado Association of Black Professional Engineers and Scientists (CABPES) and Colorado I Have a Dream Foundation (CIHADF). Within the university, highly successful organizations that serve as the foundation for the Noyce program are the Center for Advanced STEM Education (CASE) and the Colorado Wyoming Alliance for Minority Participation in STEM (CO-WY AMP). When asked what support services they had utilized at MSU Denver that had helped them be successful, 93% of the student survey respondents reported utilizing CO-WY AMP services such as tutoring. Studies demonstrate that students' experiences early in college have the most overall influences on persistence.[32] It is obvious that no simple solution will lead to higher retention and graduation rates. Instead it requires coordinated efforts to assist our students. Through the multi-pronged approached, CASE has shown a yearly retention rate with participants of 86–90% in the last three years, compared to the MSU Denver retention rate of less than 70%.
- Provide clear information about and access to financial benefits of stipend-based activities and scholarships to help these students graduate in a more timely manner.[33] When asked about the best way to access information about classes, events, and support services, more than 85% of our students said they would prefer to use a website. We have to be aware that those preferences may change as different media become available and/or more popular.
- Create supportive cohort models.[34] Students talked about how much they rely on their peers so it is only logical that we tap into what they already feel comfortable with. Our challenge at this point is that our numbers of math and science teacher candidates are so low, it's a tiny cohort. We have therefore adjusted our definition of cohort as a "community of inclusive STEM excellence" that includes students and alumni from various points in their trajectories.
- Be vigilant about power dynamics. For students who have been underprepared in their K–12 education, though our university has open admissions, students encounter some STEM faculty who treat their courses as weeding-out mechanisms. First generation to college students are from families who either had limited access to higher education and/or had negative experiences with formal education. There are also power dynamics at play in interinstitutional relationships such as between universities and community colleges. During this process, it became clear how important it was to really *listen* to their needs with power dynamics in mind.

- Recognize that inclusive pedagogy matters. Equitable STEM Teaching practices are intentionally designed to facilitate learning.[35] If we want to have different student learning outcomes, we have to problematize the status quo, White middle class approach to teaching that predominates our institutions.

Final Thoughts

The capacity building process has no endpoint. There is a spectrum of stakeholders from the funders and policy makers being the most visible, to the "targets" of capacity building who are the least visible. We need to continually examine and re-examine our institution and its practices by seeking out the voices of our students as the least visible stakeholders in our system of education. As the students and their needs change, we must change. Institutions of higher education are not generally known for being nimble. However, if we acknowledge the fact that we are here *because of the students*, we must center their voices. We need to seek out and attend to the voices of the students in our institution who feel the most marginalized. While the traditional approach has been to "weed out" such students, we must realize that they are actually the holders of the greatest funds of knowledge on the actions we need to take. Their experiences and challenges offer much insight on how to be more inclusive and welcoming educational institutions. Centering their voices is a necessary transgressive practice for helping us realize the potential transformation of formal education.

Discussion Questions

- What is the legacy of the *place* where your institution is located?
- How can you connect with the knowledge and history of peoples, communities, and ideas that were historically displaced at your site, both for curricular content and for social justice action?
- Who forms the spectrum of stakeholders in your work? What mechanisms are in place to really *listen* to your stakeholders, especially the least visible?
- How can you embed equity in the designs of new structures to meet the needs of stakeholders across the spectrum?

Notes

1. See https://msudenver.edu/media/content/equityscorecard/MSUDenverEquityScorecard2014FINAL.pdf
2. See http://www.lincolnparkneighborhood.org/history/
3. See https://cue.usc.edu/tools/the-equity-scorecard/?

4. See https://msudenver.edu/media/content/equityscorecard/MSUDenverEquityScorecard 2014FINAL.pdf
5. Howard, 2010; Moore, 2002; National Council of Teachers of Mathematics, 2012.
6. CEI, 2014, p. 16.
7. NSF Broadening Participation webinar 6/22/17.
8. See https://www.npr.org/sections/ed/2017/07/16/536488351/teachers-with-student-debt-the-struggle-the-causes-and-what-comes-next
9. See https://laprogram.colorado.edu/
10. Bryman, 2004; Denzin & Lincoln, 2008; Mauthner & Doucet, 2003; Wolcott, 2005.
11. Glaser & Straus, 1967.
12. Lawrence-Lightfoot & Davis, 1997.
13. Priest, Roberts, & Woods, 2002.
14. Lawrence-Lightfoot & Davis, 1997.
15. Johnson, Liu, Bernhardt, Evans, Koester, & Loats, 2016a.
16. Cabrera, Nora, & Castaneda, 1992; Chemers, Zurbriggen, Syed, Goza, & Bearman, 2011; House, 2014.
17. Johnson, Liu, Bernhardt, Evans, Koester, & Loats, 2016b.
18. See https://www2.ed.gov/about/offices/list/ope/trio/index.html
19. Cabrera et al., 1992; Chemers, Zurbriggen, Syed, Goza, & Bearman, 2011; Forbus, Newbold, & Mehta, 2011.
20. Cabrera et al., 1992; Forbus et al., 2011; House, 2014.
21. Otero, Pollock, McCray, & Finkelstein, 2006.
22. Chemers, et al., 2011; Maton, Hrabowski, & Schmitt, 2000.
23. Maton et al., 2000; Morrell & Carroll, 2003; National Council of Teachers of Mathematics, 2014; Semken & Freeman, 2008; Slovacek, Whittinghill, Flenoury, & Wiseman, 2012.
24. McClelland & Holland, 2015; Moore, 2002; Moss-Racusin, Dovidio, Brescoll, Graham, & Handelsman,, 2012; Tienda, 2013.
25. ACT, 2013; Gullatt & Jan, 2003; Tsui, 2007.
26. Kane, Beals, Valeau, & Johnson, 2010.
27. Hurtado, Cabrera, Lin, Arellano, & Espinosa, 2009.
28. Close, Conn, & Close, 2016.
29. Chemers et al., 2011.
30. Gullatt & Jan, 2003.
31. Patton, 2014.
32. Reason, 2009; Terenzini & Reason, 2005.
33. ACT/Council for Opportunity in Education, 2013; House, 2014.
34. Maton et al., 2000; Morrell & Carroll, 2003; National Council of Teachers of Mathematics, 2014; Semken & Freeman, 2008; Slovacek et al., 2012.
35. See the work of Paul Gorski; Christine Sleeter; Lisa Delpit; Luis Moll & Norma Gonzalez.

References

ACT. (2013). The Condition of College & Career Readiness 2013: First-Generation Students.

Bryman, A. (2004). *Quantity and Quality in Social Research.* New York: Routledge.

Cabrera, A. F., Nora, A., & Castaneda, M. B. (1992). The role of finances in the persistence process: A structural model. *Research in Higher Education, 33*(5), 571–593.

Chemers, M. M., Zurbriggen, E. L., Syed, M., Goza, B. K., & Bearman, S. (2011). The role of efficacy and identity in science career commitment among underrepresented minority students. *Journal of Social Issues, 67*(3), 469–491.

Close, E. W., Conn, J., & Close, H. G. (2016). Becoming physics people: Development of integrated physics identity through the Learning Assistant experience. *Physical Review Physics Education Research, 12*(1), 010109.

Colorado Education Initiative. (2014). *STEM Education Roadmap*. Retrieved from http://www.coloradoedinitiative.org/our-work/stem/additional-stem-information/

Denzin, N. K., & Lincoln, Y. S. (2008). *Strategies of Qualitative Inquiry*. London: Sage Publication.

Forbus, P. R., Newbold, J. J., & Mehta, S. S. (2011). First-generation university students: Motivation, academic success, and satisfaction with the university experience. *International Journal of Education Research, 6*(2), 34–55.

Glaser, B. G., & Straus, A. S. (1967). *The discovery of grounded theory: Strategies for qualitative research*. New York: Aldine de Gruyter.

Gullatt, Y., & Jan, W. (2003). *How do pre-collegiate academic outreach programs impact college-going among underrepresented students*. Washington, DC: Pathways to College Network Clearinghouse.

House, W. (2014). *Increasing college opportunity for low-income students: Promising models and a call to action*. Report, Executive Office of the President.

Howard, T. C. (2010). *Why race and culture matter in schools: Closing the achievement gap in America's classrooms*. New York: Teachers College Press.

Hurtado, S., Cabrera, N. L., Lin, M. H., Arellano, L., & Espinosa, L. L. (2009). Diversifying science: Underrepresented student experiences in structured research programs. *Research in Higher Education, 50*(2), 189–214.

Johnson, J. M., Liu, H. P., Bernhardt, P. B., Evans, B., Koester, M., & Loats, J. (2016a). [Student survey]. Unpublished raw data.

Johnson, J. M., Liu, H. P., Bernhardt, P. B., Evans, B., Koester, M., & Loats, J. (2016b). [School of Education enrollment statistics]. Unpublished raw data.

Kane, M. A., Beals, C., Valeau, E. J., & Johnson, M. J. (2010). Fostering success among traditionally underrepresented student groups: Hartnell College's approach to implementation of the math, engineering, and science achievement (MESA) program. *Community College Journal of Research and Practice, 28*(1), 17–26.

Lawrence-Lightfoot, S., & Davis, J. H. (1997). *The art and science of portraiture*. San Francisco, CA: Jossey-Bass.

Maton, K. I., Hrabowski, F. A., & Schmitt, C. L. (2000). African American college students excelling in the sciences: College and postcollege outcomes in the Meyerhoff Scholars Program. *Journal of Research in Science Teaching, 37*(7), 629–654.

Mauthner, N., & Doucet, N. (2003). Reflexive accounts and accounts of reflexivity in qualitative data analysis. *Sociology, 37*, 413–431.

McClelland, S. I., & Holland, K. J. (2015). You, me, or her leaders' perceptions of responsibility for increasing gender diversity in STEM departments. *Psychology of Women Quarterly, 39*(2), 210–225.

Metropolitan State University of Denver & Center for Urban Education (CUE). (2014). *Equity Scorecard Report for Retention and Completion*. Denver, CO: The Equity Scorecard Evidence Team and Estela Bensimon.

Moore, R. (2002). Science education and the urban achievement gap. In D. B. Lundell & J. L. Higbee (Eds.), *Exploring Urban Literacy & Developmental Education* (33–46). Minneapolis: Center for Research on Developmental Education and Urban Literacy.

Morrell, P. D., & Carroll, J. B. (2003). An extended examination of preservice elementary teachers' science teaching self-efficacy. *School Science and Mathematics, 103*(5), 246–251.

Moss-Racusin, C. A., Dovidio, J. F., Brescoll, V. L., Graham, M. J., & Handelsman, J. (2012). Science faculty's subtle gender biases favor male students. *Proceedings of the National Academy of Sciences, 109*(41), 16474–16479.

National Council of Teachers of Mathematics. (2014). *Principles to actions: Ensuring mathematical success for all*. Retrieved online from www.nctm.org.

National Council of Teachers of Mathematics. (2012). *Closing the opportunity gap in mathematics education*. Retrieved online from www.nctm.org.

National Science Foundation (2017, June 22) Broadening Participation [Webinar] http://www.informalscience.org/community/calendar/webinar-nsf-opportunities-broadening-participation-stem

Otero, V., Pollock, S., McCray, R., & Finkelstein, N. (2006). Who is responsible for preparing science teachers? *Science, 313*(5786), 445–446.

Patton, S. (2014, October 27). Black man in the lab. *The Chronicle of Higher Education*, Diversity in Academe Report.

Priest, H., Roberts, P., & Woods, L. (2002). An overview of the three different approaches to the interpretation of qualitative data. Part 1: Theoretical issues. *Nurse Researcher, 10*(1), 30–42.

Reason, R. D. (2009). An examination of persistence research through the lens of a comprehensive conceptual framework. *Journal of College Student Development, 50*, 659–682.

Semken, S., & Freeman, C. B. (2008). Sense of place in the practice and assessment of place-based science teaching. *Science Education, 92*(6), 1042–1057.

Slovacek, S., Whittinghill, J., Flenoury, L., & Wiseman, D. (2012). Promoting minority success in the sciences: The minority opportunities in research programs at CSULA. *Journal of Research in Science Teaching, 49*(2), 199–217.

Terenzini, P. T., & Reason, R. D. (2005). *Parsing the first year of college: Rethinking the effects of college on students*. Paper presented at the Annual Conference of the Association for the Study of Higher Education, Philadelphia, PA.

Tienda, M. (2013). Diversity ≠ inclusion: Promoting integration in higher education. *Educational Researcher, 42*(9), 467–475.

Tsui, L. (2007). Effective strategies to increase diversity in STEM fields: A review of the research literature. *The Journal of Negro Education, 76*(4), 555–581.

Wolcott, H. F. (2005). *The art of fieldwork* (2nd ed.). Lanham, MD: Altamira Press.

Implications and Conclusions

JOY BARNES-JOHNSON[1] AND JANELLE M. JOHNSON[2]

Editorial Reflections

Unlike in previous chapters where we share our stories as a means of previewing the chapter that follows, these reflections simply represent our "exhale" as we contemplate the sum of each part presented here. Each contributor to this volume shed light on the challenges we face as vested STEM actors. We recognize the urgency that is needed to respond to these challenges, especially in STEM education because it is proving itself to be "...a [critically] important element in the struggle for human rights":[1] educación es equidad! There is so much that we have attempted to say on these pages to help us collectively understand what EQUITY in STEM education means for the 21st century. We hope

---

1. Joy Barnes-Johnson, Science Educator
   Princeton (NJ) Public Schools
   joybarnesjohnson@princetonk12.org
2. Janelle M. Johnson, Assistant Professor of Secondary Education
   Metropolitan State University of Denver
   jjohn428@msudenver.edu

we got it "right," and that you find these chapters and stories useful in whatever context you find yourself.

This volume is for STEM stakeholders en masse: industrial partners, policy makers, higher education, compulsory education and community educators. We purposefully presented and interrogated the systems and structures that oppose socially just and anti-racist, anti-sexist approaches to STEM, across academic and professional contexts. We recognize the magnitude of this task. We understand that transgressive practices can be uncomfortable and discouraging, especially when the work calls for building racial knowledge. Western viewpoints[2] and the established boundaries[3] of Whiteness cultivate deficit thinking. When we envisage the normalization of Whiteness as a widely held disposition of White teachers—the majority of the education workforce from prekindergarten through higher education—we resist being discouraged because White actors are not the only ones responsible for this problem. Whiteness and Western viewpoints are dominant and therefore dominate. We experienced this in real time as we went back and forth during the editing process. For us, two careful fans of critical pedagogy, we recognized when we were succumbing to the pressures to use standard conventions of language and intellectual etiquette. This work is tough ... we know that.

"Right now, dominant [academics] overshadow critical [learning]. The goal is parity between them."[4] This is what Gutiérrez (2002) described as "enabling equity practice" which sheds new light on the idea of "transgressive practices" in and for 21st century challenges. People, however, need a "safe" place to grapple with their own complicity, cognitive dissonances, and habits of mind that generate oppression, so we ask ourselves: have we done that? Have we created a safe space to think about equity and social justice in STEM? We, the editors, understand that the types of dismantling/change we imagine is by nature disruptive, confrontational and often uncomfortable. We have shared our perspectives and posed questions in an effort to make sense of injustice in STEM as we have learned to be comfortable with our own discomfort. We, Janelle and Joy, are friends—unafraid to ask questions of each other or hold each other accountable for our own bias and so, we have invited you, the readers of this project, into our own process.

We met in 2014 because a colleague connected our work after attending a STEM education conference. This small network of three talked about the wonderful potential of equitable science teaching (EST) and the Next Generation Science Standards (NGSS) to change the state of STEM education in the Rocky Mountain region. Like so many times before, we were coming out of the shadows of the speakers who had presented research findings at STEM conferences and were building alliances and networks with people who were open to asking and being asked questions. We, like many others at large conferences, could be found sitting silently, questioning the work of the field—silent perhaps because we were "early career" academics whose views were informed by our experiences as students and practitioners to a much greater extent than any habits or customs as experts. Whenever we meet, we often talk about the presentations we've heard or things we've read and what is missing from or understated within them. We generally always draw the same conclusion: the voices of those from non-dominant cultures are missing. In STEM, non-dominant cultural groups are not merely defined by class, race, gender or ethnic categories but by able-ness, able-bodiedness, nationality, and faith/spiritual practices (among many others). What we imagined over a cup of coffee at the conclusion of yet another STEM education conference in 2016 was a volume that took a woman who looked like a "Becky" and a woman who looked like a "Khadijah" and honored what we brought (and could assemble) to the table as thinkers.

Opportunity: Positioning Equity as Transgression

Throughout the volume, the deficit-based "achievement gap" that locates blame on students and families has been reframed as an opportunity gap, inviting STEM providers to engage in praxis, reflecting on possible changes they can make to their own structures and in their staff. The students who are described as "lagging" are most often the students whose educational experiences are represented nearly all by windows rather than mirrors or doors.[5] They are students who learn about innovators who do not look or sound like them, instead their *seeing* is as gaze, *aspirational* as it may be, through a divider that separates their reality from other normalized lives. They are students whose own families, if they had access to school, likely faced much adversity themselves and may hold a negative view of schools and teachers. Their teachers do not look like them; there are few (if any) mirrors for them to find *inspirational* on the pursuit toward school-specific learning goals. If they make it to college, even fewer of their professors will look like them. They will stand out in their classes. They will feel isolated. These students need opportunities to learn (OTL) throughout their cradle to career trajectories; that work is on all of us. We hope the chapters of this volume have served as a *granito de arena*[6] to this transgressive border-crossing effort, and we share this chapter in conclusion. We begin by revisiting how the volume's journey began.

Project Overview

The number of research publications critically examining STEM equity—with depth of understanding and intentionality—is surprisingly miniscule, in spite of the urgency to address it. As a reader and consumer of information, you may find the same keywords in lists or searches within the text however the complexity, unexamined biases and hidden meanings buried within texts may not always be clear. When we started this volume, we did not know or intend for it to be its own grounded theory[7] work. We simply sought to create a space where researchers and practitioners could collaborate and share ideas in an honest forum. We had hoped to cull the resources on STEM education and prioritize issues that we cared about: equity and inclusive pedagogies. As editors of this volume, we invited the research community to a conversation from the perspective of practitioners.

Call for Proposals—October 2016
The editors of this volume seek to provide the research community with perspectives on practices that effectively engage underrepresented communities in STEM teaching and learning.
The editors are seeking chapters based on grounded theory research, program evaluation, policy and action related to the Next Generation Science Standards and/or the Every Student Succeeds Act of 2015. Contributions may include but are not limited to the following topics:

- *Equitable STEM Teaching*
- *Transdisciplinary Connections with Common Core Subjects (Language Arts and Mathematics)*
- *Preparation and Professional Development of Teachers for Work with Multilingual Students and Families*
- *Engagement of Students from Underrepresented Communities*
- *Pedagogical Strategies for Inclusive STEM Engagement*

Please send a 500–1000 word letter of intent including abstract by December 23, 2016

Using contact lists, social media platforms, professional and academic listservs to distribute the call, we initially got fewer than fifteen responses. From twelve chapter proposals, three of which represented our own work, we selected nine "reports" for inclusion in this volume. Viewing these initial submissions as research applied to practice, we collated findings from researchers—many of whom were graduate[8] students—grappling with equity as a theoretical paradigm in STEM education at a time of significant and yet unknown sociopolitical change. Clear lines of discourse emerged that threaded each paper to the other. Even when authors used challenging terms and phrases to convey their work, we as editors knew that troubling language around equity is perhaps one of the major reasons why it has not taken a more prominent place in the canon of high quality reports of best practice in STEM education. It is hard to argue that equity is not important, but equity can be challenging to explain outside of very specific contexts. Many of the practitioners and researchers who may be willing to raise the issue don't really *want* to navigate waters around "antiracist" education, especially in the current sociopolitical climate of banned terms and dismantled civil rights webpages. We dive into those waters with these two primary questions from the book's introduction:

1. How does this STEM experience (teaching and learning) promote transgressive thinking?
2. What do *we* need to do differently as STEM education providers, as teaching and learning facilitators and as STEM education researchers?

Based on the work of the authors in this volume, we have identified very specific indicators and characteristics (observable attributes) of equity in research that inform practice. We made transparent our own challenges as STEM teachers and learners by sharing our stories in the editorial reflections: our unique reactions to the contributors' works are reflexive and somewhat iterative—across the pages, we are in dialogue with the contributors, posing questions with and to them. Working from practice to theory, we synthesized our understandings of the narratives being told by the contributing authors from the standpoints of our own experiences. We feel that our most interesting "finding," discussed later in the chapter, is the need for a fourth dimension to the current three-dimensional framework for K–12 science education[9]—equity. The proposal of a fourth dimension is not new. In an analysis and critique of the NGSS for example, *nature of science* content—a set of traditionally centralized ideas about how to do and act as a scientist—was suggested as a missing dimension.[10] Like nature of science, our proposal to include equity as a fourth dimension to the framework is not new either. *Equity* is among the other appendices (including nature of science) to the Next Generation Science Standards that emerged from the framework. However, it is not centralized or prioritized[11] in a way someone new to STEM equity work would find useful, a point documented by scholars like Alberto J. Rodriguez and Jomo Mutegi. Present performance expectations, learning outcomes based in the three dimensions[12] appear objective and therefore attainable but are not often easy to do or well executed. Novice and veteran teachers still struggle to build fully participatory learning environments where all students benefit and/or find the content useful; researchers struggle with how to capture episodes of NGSS implementation success. It is not difficult to understand some reasons why: lagtime between theory and practice. When STEM majors are prepared to be professional or academic leaders through traditional lecture methods and isolated cultural norms of merit, how or when will they learn to operate outside of those habits? Performance expectations that have not attended to the actors' social, cultural and emotional needs will likely never yield broad participation by a diversity of actors. Equity is the principle that transcends the normal and creates opportunity for the imagined.

Defining STEM and STEM education in the context of equity is a critically important exercise for the research community. There are many examples of teacher resources intended to address equity in development;[13] a quick search points to them directly. However, the special demands, needs and interests of a society increasingly dependent on technology[14] have shaped new discussions; technology designers have become the new policy makers.[15] The "familiar" differences associated with gender,[16] racial, ethnic, class and linguistic diversity are now coupled to dynamics associated with a growing school population formed by refugees and the children of undocumented workers. Contemporary multicultural education must therefore be re-defined; various technologies bring these data formerly difficult to identify forward for public scrutiny. The language we use to scrutinize education can establish and sustain the course or confuse it.

Section One of the volume examines key language used to outline the STEM discourse with a critical theoretical lens. Barnes-Johnson describes the historical context for thinking about equity in STEM/Education in Chapter One. By not being precise in our use of language, we may have perhaps left too much up to the sociological imagination, leaving too much room for misconceptions, mal-intent, and misuse. Building on these ideas in Chapter Two, Adjapong offers a framework for teaching and learning anchored in the creative elements of Hip-Hop. His exposition of communication styles and behavioral practices of students and teachers focuses on developing and identifying culturally relevant approaches to teaching STEM to all students, specifically urban youth. The chapter analyzes the ways equitable science teaching (EST) as a construct provides researchers with a context for understanding the beliefs and practices, the choices and the responses that teachers make in light of the various differences they face in the classroom. This has been true for more than sixty years! History shows us the hopes of previous generations: the optimistic belief that access to science and technology in schools (i.e., early exposure) would build national pipelines for adult engagement and interest in economic competitiveness. EST engages politicians, STEM industry leaders, teachers and students in real-time interactivity. In Chapter Three, the social justice themes needed to advance the idea that "equity in STEM" equates to legislative investments[17] in workforce, community, economic innovation and healthcare were explored by Asante et al. using hydroponics in out-of-school time learning as the tool, fighting the fight against environmental racism. As a society, our thinking about STEM has evolved from patriotic attention to STEM (a 20th century model of EST) built on space exploration. We are now challenged to think about systems of power that create barriers to full access to STEM participation and education, like technology and resources to live or the impact of trauma[18] on individuals and communities as they try to rebound from unimaginable stress.

In Section Two, projects that exposed youth to authentic research settings are reported. Realizing that extended learning opportunities are the ideal way to prepare students to interact in the larger universe of STEM beyond classroom constraints, the challenges of sometimes contrived "stages" are real. In Chapter Four, Keiler and Robbins provide insightful perspectives on the use of peers in learning environments, disrupting school-based systems of pairing that often reinforce negative learning and social dynamics. True 21st century EST diminishes power dynamics that position certain knowledges over others, while also allowing students to develop skills that prepare them for a range of different knowledges. Borrowing from Gutiérrez's (2002) equity framework, the goal of providing a mechanism for the meaningful contributions of marginalized students to participate in the "erasure of inequity on the planet" as problem finders, solvers and producers of change will require deliberately focused effort of student mentorship. In Chapter Five, Xu, Newton, Turrin and Vincent share snapshots of the sometimes-awkward interactions between the STEM community and the education community. Teaching humans STEM often means teaching humans about STEM epistemologies and sociocultural hierarchies, pinning survival in STEM ecosystems to that understanding. In Chapter Six, Bhaduri et al. expand the notions that engineering and play go hand-in-hand, ideas first advanced in the 1960s by STEM pedagogy giants like Robert Karplus.[19]

Section Three revisits our theoretical framework of transgressive practices. Chapter Seven provides samples from actual teaching and learning episodes that integrate STEM with other content areas. In Chapter Eight, Suess, Chae and Lewis look at community health and community/public health education and the historical tension that exists between those who "serve" and receive services. The legacy of distrust persists like justified folklore crafted in microclimates that are hard to deny. Chapter Nine extends this narrative; tension between groups and the agency, or lack of agency, that comes as a result of these tensions is described by Johnson, Schendel, McClellan Ribble and Liu. Arriving at the conclusion that inclusive pedagogy leads to capacity is one that is not hard to understand, though it is difficult to achieve. Every type of STEM literacy is impacted by this need for change; all depend on new thinking about how to embed social justice[20] in STEM projects and programs.

Building on the seminal works in multicultural education done by Professor James Banks of the University of Washington, social justice approaches to learning contribute to empowering school culture. These are easily understood to be the most highly evolved forms of multicultural education among its five dimensions: content integration, knowledge construction, equity pedagogy, prejudice reduction and empowerment.[21] Efforts made by the authors of this volume, *STEM 21: Equity in Teaching and Learning to Meet Global Challenges of Standards, Engagement and Transformation*, to build stakeholders' confidence, self-efficacy beliefs and esteem characteristics have been considerable. As stakeholders develop beliefs about their own ability to think and do as scientists do, their capacity and ability to do science increases. Realizing that science and engineering practices are normal human enterprises like art and communication is part of that esteem-building framework. Fully aware of the power of language to transform students' learning experiences, aspects of literacy, the asterisked R (R*) in STR*EAM combines many literacy and justice ideas: reading, rhyming, [w]riting, righting,[22] reasoning and research. Each is positioned within the equitable STEAM teaching and learning context as an opportunity to express voice, build agency, and realize "enabling practices" that develop proficiency in dominant culture content, develop critical stances and new perspectives in STEM while contributing toward the positive relationships of STEM knowledge producers and STEM product consumers.[23]

On Heroes, Mentors and Change Agency: New Insights about Representation and Participation as Viewed Through the Theoretical Framework

Transgressive education engages, excites and leads to self-actualized learning where self-expression, and spiritual growth and wellness are possible (hooks, 1994). How do we then re-form STEM education into a more transgressive form? Even when clumsy or asynchronous, full participation leads to powerfully enabling equitable practices.[24] Coming to the table of equity when the loud chatter of overly active, excited youth disrupts the familiar silence of the academy is difficult; the cadence of Hip-Hop in traditional STEM environments is "tolerated." Yet, these episodes often leave us transformed by a simple question: how can we, the established STEM actors, find the same kind of joy that these youth wield so effortlessly? Students *do* need to see themselves in the curriculum in order to connect with it. Borrowing from the ideas presented in Emily Jane Style's "Curriculum as Window and Mirror,"[25] we recognize that all learners need mirrors to see themselves and connect with the curriculum, to be able to imagine STEM pathways for themselves, and to find the strength to persist in the face of the challenges they will most certainly encounter. Learners also need windows to gain knowledge of communities other than their own. For students who come to school with more privilege, they typically see lots of mirrors—their own cultures, values, and community members are reflected but they lack knowledge and understanding of the students who come to school with less privilege, whether based on race, socioeconomic status, language literacy, able-ness, or legal status. This void is being inflamed in the current political climate, with various bodies actively working to pit one group against another. If we can help teachers and students develop windows into cultures other than their own, we believe more empathy will result, and more critically aware folks will be less likely to fall prey to calculated divisiveness. This can result in more diverse, more productive STEM learning across the cradle to career continuum, moving towards the dual goals of social justice and improved academic outcomes for all.

Representation

Capitalizing on the necessity of more equitable representation in the delivery of anti-racist/anti-sexist experiences for STEM students, STEM educators and change agents have a responsibility to encourage students to really think about the histories and legacies of heroes and communities where marginalized students are. When students are assigned or elect to study "ordinary and extraordinary"[26] historical figures or events in the context of social justice in STEM,[27] they are themselves embracing multicultural education's "first step." Whether students learn about the contributions of a diversity of actors, see models worth emulating, or base their efforts on that of other STEMJ workers, they are moving closer toward transgressive practices on their own path to achievement.

Mentorship is Leadership

As they realize their own aspirations, students develop a sense of belonging to the broader STEM community. By listening to the cues youth provide[28] about why they lack interest or what they really want in an experience, opportunities are created to evolve and adapt programs to be more meaningful and enjoyable for everyone. Types of mentorship valued most by youth include near-peer and direct-peer mentorship perhaps to a much greater extent than adult stakeholders know. The social implications of direct-peer and near-peer mentorship, make this type of mentorship a unique opportunity to forge relationships, build networks and experiment with both the risks and rewards of social engagement in STEM learning environments and cultures. Similarly, as students contextualize the contributions made by heroes more broadly, new insights about social justice, the role of STEM in societal context and the need for broad participation in STEM across cultural divides becomes apparent.[29] Problem-based learning centered on important issues like food (in)security and Tribal nations' demand for mineral rights protection can be used to drive this point home. Everyday scientists and citizens who take on these problems are STEM heroes; they are using STEM knowledge and profound understanding of STEM and community to build learners' agency and capital.

Media Influence

Beyond the traditional canons from which we pull references to heroe-dom, STEM can also look to science fiction and animated forms to locate examples of excellence; somehow the artistic world has captured and represented the critical, watershed moments of real life that feed new interests. STEM-related films like "Back to the Future," "Terminator," "Mad Max," "Star Wars," "Star Trek," and "The Martian" have few (if any) Black or Latinx characters. Ryan Coogler's "Black Panther," a 2018 blockbuster film, has taken this void and challenged viewers not only to see Black STEM agency, but to understand that community knowledges are valued in concert with contemporary technologies, and that youth voice is central to building capital. While it may seem peculiar that comics are the vehicle being used to point out racist practices underlying much of STEM, the art-imitating-life dimensions of fantasy narratives are transgressive.

> African American heroes are … defenders of Earth … the villains are racism, classism, sexism and poverty, most of which are inflicted by non-African Americans, and legislated by our own government. Where are the [real-life] heroes addressing the same issues that exist in comics? African American heroes can be a conscience of sorts for comics. Our heroes don't wear masks. Why? Simply put, our heroes represent realistic ideals. Justice, freedom and equality don't wear masks. (Stacey Robinson as cited in Duffy & Jennings, 2010, p. 163)

Assigning new authority to literature and media as STR*EAM pedagogy moves into the toolkit of best practices for the 21st century is appropriate. Science fiction has always inspired many, but as more multicultural casts take prominent roles in portraying and delivering media, we can expect an even broader and deeper impact. We celebrate Ava Duvernay's 2018 "Wrinkle in Time" for its STEM themes and content, respect for Indigenous and spiritual knowledges, as well as its diverse cast. The creators and illustrators of these fictional characters are amplifying this truth.

Real-world STEM Heroes: Past and Present Commitment to Social Justice

Little-known and taken-for-granted heroes of real-life STEM include Charles Henry Turner,[30] Susan La Flesche Picotte,[31] Jewell Plummer Cobb,[32] Maryam Mirzakhani,[33] Percy Julian,[34] Vivien Thomas[35] and Leona Marshall Libby.[36] They are as important for us to know as Alfred Noble, Albert Einstein, Rachel Carson or Marie Curie. Who is Erin Jackson[37] or Princess Shuri of Wakanda?[38] Each is a pioneer who has opened doors that were generally shut to them. Their stories in historical context could easily serve as models of courage to do STEM in socially just ways. If all students are invited to study the life and achievements of STEM leaders from within non-dominant cultural groups, it serve as an impetus to help students know that they too can be agents of change for improved STEM learning, socially just thinking and high productivity as global citizens. This challenge is not to help institutions meet "diversity goals," but rather to centralize the needs of its stakeholders. There is cause for hope because some progress is apparent, but this work is never done. We must move beyond trivial "tolerance" of otherness (or what amounts to the exploitation of available populations) in order to meet superficial broader impact goals, a fallacy of tokenism that we must resist. In order to genuinely transform our work, the questions we ask must be compelling, not just for serving the dominant view, but creating opportunities for formerly and presently silenced voices to have a place at the table.

Voice and Agency in the Process of STEM

In the 21st century universe of STEM understanding, the function of science, technology, engineering and mathematics is uncertain. Challenged by current definitions of STEM, and banned words lists, the current generation of people doing, seeking to do and teach STEM is talking about it. According to a recent thread on Twitter, @zerdeve, a stat/psych person, offered this perspective: "My very honest but naïve opinion about science today: we somehow stopped valuing systematic observation and naturalistic approach[es] to science and started thinking we can keep testing hypotheses instead … but we're not spending enough time developing hypotheses worth testing." Surprised by the responses from others, this user's initial thoughts launched discourse that was both arresting to those doing STEM and those observing its impact from various vantage points. Ultimately, scientific methodologies lead to engagement in STEM as initiated by observations of problems or as a response to various proposed solutions. Historically, we have known perhaps *what* each discipline means and/or does but we have been less careful to actually hold people accountable to the refined purposes of these disciplines to serve more than individual gains. We think about meritocracy and democracy as separate, yet in most social, economic and political settings, achievements are rewarded in socially, economically or politically unbalanced ways—the very definition of inequity.

Conclusions—Equity as the Fourth Dimension in STEM Education

The scale of global problems we face have magnified the consequential realities of STEM ignorance; the general population is able to discern the many shifts in collective thought about science, technology, engineering, and math, highlighting the need for better approaches to teaching and learning. The Next Generation Science Standards and their underlying three-dimensional framework offer important shifts toward interdisciplinarity, yet equity is on the periphery of their design. While this fact may reveal their authors' positionality, it is an Achilles heel that limits implementation with the diverse learners they purport to serve.

The United States seems to be falling behind all other countries in making progress toward solving environmental, social and technical issues because of cognitive dissonances created by political chaos and inhumane algorithmic musings.[39] The challenges are "difficult and exciting ... both social and technical" (Lee, 2016). Building on a legacy of multicultural education and culturally relevant education and policies that support access to quality education for all, this is a watershed moment for equitable STEM education. Threats to the ways we do STEM education research and operationalize language around STEM education have placed researchers and practitioners in close intellectual proximity. With the arrival of video resources like iTunesU and TED Talks, acceptable notions of highbrow conference proceedings, ethereal closed lectures and inaccessible disciplinary jargon have been completely disrupted. The TED organization itself remains one of the more interesting portals for knowledge sharing between industry, the academy and the populus almost twenty-five years after its first convention. Established in 1984, the TED talk, an acronym representing the convergence of technology, entertainment and design, uses web-based videos to share conference insights from keynote speakers embracing current standards of STEM best practice: robust methodologies, accuracy, transparency and rigor.[40] Its attendees (its viewers) are made privy to profoundly impactful research in a wide range of disciplinary fields. Some of the most popular talks address issues like the neuroscience of failure (Brené Brown, 2010/2012); urban gardening, public health & food deserts in Los Angeles (Ron Finley, 2013); Indigenomics showcasing the need to include Indigenous populations in genomic research using portable nanotechnologies (Keolu Fox, 2016) and the healthy distrust of algorithms created from "big data" (Cathy O'Neil, 2017). Unlike previous generations where information could only be culled from books or hard-to-get periodicals, social media platforms give the current generation of thinkers access to resources that has created conditions prime for us as a society and subcultures within it to grapple with compelling questions about intellectual property, ethical capture and use of data and crowdsourced information. This combination of questions, evidence from vetted research and transparent engagement with the public has created new demands on STEM that must be addressed. The public now contributes freely to STEM discourse and demands publicly that STEM actors have integrity and remain lucid. This is the perfect climate for considering social justice and equity as a fundamental fourth dimension of STEM.

Social Justice Approaches to STEM

In closing this volume, we present what emerged as a theoretical anchor for equitable STEM education. Social justice is ultimately a tool used to celebrate diversity and resist practices of EXclusion.[41] It interrogates the systems and mechanics of oppression and/or discrimination with acute attention to non-dominant groups. In a social justice approach to STEM (STEMJ), research that supports understanding the human conditions that produce inequity in natural (e.g., ecological communities) and manufactured spaces (e.g., labs and classrooms) are prioritized over capitalistic studies that exclude or downplay the human experience. This approach represents transgressive practices, a torch lit by bell hooks in the early 1990s, that we are using to illuminate an inclusive pathway forward in STEM. bell hooks' framework of transgressive practices allows us to apply the standards of engaged pedagogy and liberatory practice to different learning contexts. We have treated each paper as data, extracting from the various literature perspectives and research findings presented similar to qualitative methodologies of coding based on themes,[42] analyzing it for particular indicators of STEM equity. We created codes that we are calling "subtexts," and worked to articulate specific characteristics for each, described in the table below. Attributes of each equity subtext are drawn from the chapters in this volume to show how researchers transgress toward STEMJ, supporting the greater purpose of taking positive actions to eliminate discrimination and oppression in STEM teaching and learning environments.

Table 10.1: Descriptions of STEM Equity Indicators (by subtext).

Subtexts	Attributes
Social justice (SJ)	Food security; public/community health; responsible land use; respectful land acquisition
Mentorship (M)	Near-peer; direct-peer; adult-student; scientist-citizen; industrial-academic; relationship building; experts as advocates
Student agency (SA)	High interest; relevance for personal sense making; voice; out-of-school/informal; flexible time/space orientation; fosters/supports heutagogy;[43] builds communities of practice; provides pathways for social learning/network building
Teacher agency (TA)	High interest; relevance for personal sense making; voice; out-of-school/informal; flexible time/space orientation; fosters/supports heutagogy; builds communities of practice; provides pathways for social learning/network building; professionalization of teachers/ing

(*Continued*)

Table 10.1: (*Continued*)

Subtexts	Attributes
Student inquiry (SI)	Shift from consumers to producers of knowledge/resources/tools; sense-making; student directed; solves real-world problems
Interdisciplinarity (I)	STEAM/STREAM; multiple science disciplines; cross-curricular; real-life contexts; career pathways
Place-based education (PBE)	Community based; self-determination; social-economic development pathways; intracultural; intercultural; fight for mineral rights[44] or water justice[45]
STEM Ecosystem Health (ECO)	Out-of-school time; social interaction within and between agents and/or systems; similar to mutual or commensal relationships, parts of a healthy STEM education ecosystem are not harmed within the system; healthy STEM ecosystems operate where individuals (or groups) operating within it understand that there may be no [immediate] benefit for one actor and the benefit(s) to other actor(s) may be delayed or long-term
Standards-based (SB)	Common core, Next Generation Science; develops teacher dispositions and competencies for diversity; differentiated instruction; high expectations and higher order thinking for all
Empowering identity formation (ID)	Self-efficacy; collaboration within and outside of STEM education; identity-independent agency. Representation. Fictional examples: Doc McStuffins[46] and Princess Shuri and Star Wars (issues of justice) Non-fiction: "heroes" and contributions

Source: Authors.

Tables 10.2 and 10.3 locate these subtexts as STEM equity indicators in the volume's chapters, indicated with shortened titles for readability. The first table examines each chapter's subtext in terms of a *teacher* focus. The second table focuses on *students*. The system of notation indicates whether the subtext was a primary or secondary aim of the chapter. Asterisks are used to point out more traditional and potentially non-transgressive practices. As you have seen throughout the volume, equitable STEM work is complex; there is always room for growth. All of the authors in this volume have taken a risk by sharing their efforts with the world. This has been a learning experience for all of us. As editors, we hope the dialogue continues.

IMPLICATIONS AND CONCLUSIONS | 239

Table 10.2: Equity—The 4th Dimension of the 3D Framework of STEM Education.

Teacher focus

Chapter (Abbreviated Title)	STEM Domain				Teacher agency	Social Justice	Interdisciplinarity	Place-based	Standards-based	Empowering Identity	STEM Ecosystem Health
	S	T	E	M							
1: LEASE	X	X		X		X	X		X		X
2: Hip-Hop	X					X	X	X			
3: Seeding	X		X		X	X	X	X			X
4: New Roles	X				*				X	X	X
5: Early Engagement	X				Y	X		*	X	Y	X
6: Middle School Motivation		X	X			X	X		X		X
7: Engaged interdisciplinary literacy	X	X	X	X		X	X		X	X	
8: Transformation	X				X	X		X		X	X
9: Capacity Building	X			X		X		X	X		X

Source: Authors.

Table 10.3: Equity—The 4th Dimension of the 3D Framework of STEM Education.

Student focus

Chapter (Abbreviated Title)	STEM Domain				Student agency	Social Justice	Mentor- ship	Student inquiry	Inter- discipli- narity	Empow- ering Identity	STEM Ecosystem Health
	S	T	E	M							
1: LEASE	X	X		X					X	X	X
2: Hip-Hop	X				X	X	X		X	X	
3: Seeding	X		X		X	X	X	X	X	X	Y
4: New Roles	X				X		X	*		X	
5: Early Engagement	X				X		X[47]	X		Y	X
6: Middle School Motivation		X	X		X	X			X		X
7: Engaged interdisciplinary Literacy	X	X	X	X	X	X		X	X	X	
8: Transformation	X				X	X	X			X	X
9: Capacity Building	X			X	X	X	X				X

Source: Authors.
X: Reflected in the research design/methods presented, Primary focus
Y: Reflected in the research design/methods presented, Secondary focus
*: Traditional/surface level (reinforces potentially traditional/non-transgressive practices)

IMPLICATIONS AND CONCLUSIONS | 241

Final Thoughts

One important way to advance transformative dialogue in support of STEM equity is to seek out the voices of the foot-soldiers doing the work—the practitioners, who are largely missing from the discourse; the elevated positions of tenured or well-funded researchers in the field takes precedence over more humble projects at a national level, both public and private. As a result, the status quo in STEM education is maintained, and a history of "who said it loudest" leaves silent and invisible the works of those "who see it first." We became editors of this volume because we bore witness to the subordinated positions of students and STEM equity research "workers" in the context of elitist traditions and norms of behavior within the academy, in organizations, on buses, and public spaces. We interrogate the nature of annoyance in the dominant culture. Things like noisy classrooms, handwritten assignments, intricate doodles, playful productive interactions between stakeholders, curious musings and slightly tangential questions are acceptable by-products of engaged STEM learning—these so-called annoyances are often significant evidence of interest, motivation, and aspirations to be a part of an authentic community but written off as meaningless, disrespectful,

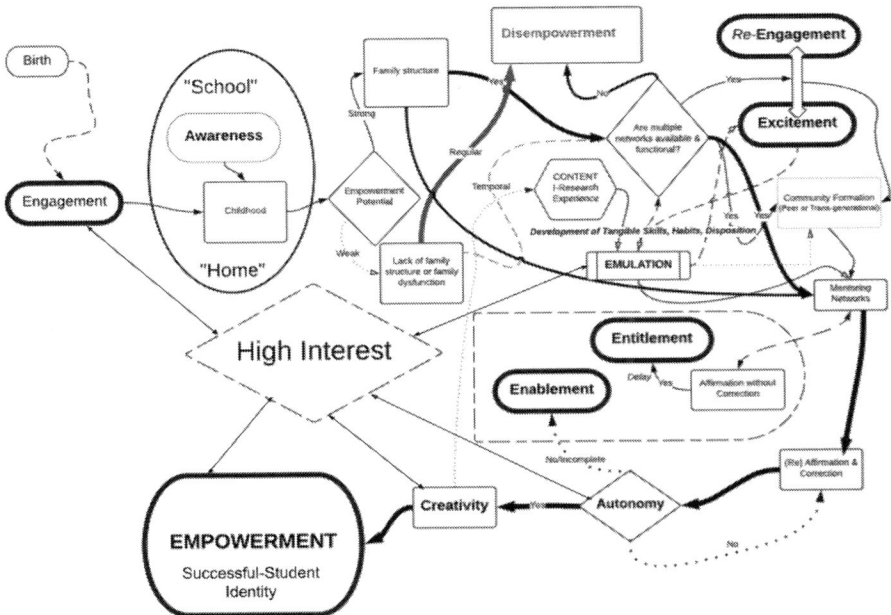

Figure 10.1: Network Theory of Successful STEM[48] Student Identity Development.
Source: Barnes-Johnson.

off-track, wasteful or punishable horseplay. If we hope to grow communities of practice that have the capacity to enact equitable STEM education, we owe it to ourselves to look beyond the obvious. Similar to discourses that problematize the singular channel *pipeline* analogy, we posit that there are multiple pathways to success in STEM education that are deeply embedded in networks of social interaction and less normative ways of being—epistemologies—that celebrate diversity in socioculturally authentic ways. These pathways are mapped out as tributaries and branches and loops and gateways like a watershed. There is no linear conduit or well organized checklist to follow, only access points and optimal conditions. As facilitators, we have the potential to reduce barriers, smooth transitions, make transparent formerly hidden entrypoints and make more permeable the veils that separate STEM agents from others.

Note: An open rather than hierarchical network of ideas, people, events and structures can either support or serve as barriers to student success. Initially drafted as a heuristic to explain empowerment pedagogy for scholar identity development[49] in Black males, this updated version of the schematic below exposes threats to success narratives and theorizes about how developmental "delays" can occur. When applied to STEM agency, high interest in STEM-related topics and activities support two-way exchanges between student actors and the various terminal points. Perforated borders or loosely tied connectors are more easily crossed than solid ones. Opportunities to engage and learn STEM concepts, emulate the actions of those doing STEM, do STEM research, create STEM products and find success while doing STEM activities in formal or informal ways make 'aha' moments typical and may lead to the development of an identity as a STEM "person": someone who is able to think and act and be a scientist.

Understanding the pathways of movement from birth to *empowered* is not as complex as the model might suggest: it is designed to help explain how STEM stakeholders can resist and counter the "overwhelming boredom, uninterest, and apathy that so often characterizes the way [educators] and students feel about teaching and learning" (hooks, 1994, p. 10). The network is open, flexible, and has limited hierarchical organization, a deviation from the traditional apprenticeship model. The teacher is positioned off-pedestal and on-par with the learner and the learning environment. Students learn to emulate best practices of mentors while avoiding the patriarchal loops and missteps of simply reproducing what mentors have done. Students' voices are not silenced in this model but there is full acknowledgment that youth need guidance, support, and adult advocates along the way. According to bell hooks, "commitment to engaged pedagogy carries with it the willingness to be responsible, not to pretend that [educators] do not have the power to change the direction of our students' lives" (p. 206). Entitlement-enablement loops that trap youth are often created by apathetic adult tendencies, avoidance of correction, ignorance or arrogance; when appropriate networks exist, mutual trust between youth and adults is more possible, appropriate crediting occurs, and youth are able to motivate themselves to exit the loop. *This* is the site of transgressive practice and increased potential for identity development as a successful student in any learning context, and certainly in STEM.

Humans are naturally curious and able to make observations based on sensory experiences that convert initial awareness into opportunities for the development of *high interest* over time; this is centered in the schematic. The first pathways to engagement in scientific practice come from the sensory experiences associated with the physical act of being born; these represent movement toward an awareness of the natural world, without regard to ability or ablebodiedness. The potential/likelihood of moving from birth toward forming an empowered identity supportive of STEM participation is then shaped by the degree of supports in school, in the home and by opportunities to learn: the richer the experiences a student has with a subject, the more likely it is that tangential touch points become secured anchors capable of connecting them to activities in stronger, less temporal ways; students see themselves through a balance created by careful placement of more mirrors in their everyday life—their life curriculum—than windows.

Building relationships increases the potential for continued movement toward empowerment. The common mistake made is to ignore the power of strong relationship and gaining entry to families and communities. Research communities often violate sacred codes of communication, manipulate the researched or undervalue that which exists outside of dominant knowledge stores. In these contexts, some still try to become partners, which may be why gaps caused by distrust persist. Observed in the downward arrow from "potential" in the schematic, the consequence of sustained and regular weak partnerships with family (either through rejection of community knowledges or limited involvement) may be disempowerment, often leading to disengagement from STEM altogether. A simple explanation of this problem rests in the "leaky pipeline" hypothesis.[50] Other important networks exist that can lead to re-engagement. The nets[51] created by mentoring relationships inside schools and other settings (e.g., churches, clubs) whether formal or informal, provide opportunity for peer and/or intergenerational transfer and growth. The influences of others and the collective efficacy-building that comes from strong and cypher-like affiliations breeds emulation. Production of shared language and ideas create knowledge bases that are cultural but also transferable. This development of tangible skills, habits and dispositions allow actors to revisit the mentoring spaces, try out their ideas with guided supports, and push past a dependence on the mentor space for independent action as a STEM native. The preparation that comes from spending time in a research environment is invaluable to skills development. Described here as "Content I-Research Experience," independent and/or immersive research experiences provide authentic contexts to learn deeply. Seen throughout the volume, this kind of immersion is possible in a school environment (e.g., a research class or intensive project), home space (e.g. family garden), community space (e.g. structured club or service group) and/or institution of higher learning (e.g. facilitated research): they all provide academically rich opportunities to learn, to act, and to build STEM/STEMJ capacity for both individuals and organizations. The affirmation and correction that occurs as a result of iterative mentoring when accepted and processed well makes room for co-authorship, autonomy and creativity. If the actor rejects these elements, experiences failure (being rejected outside the safety net or is disaffirmed), or is affirmed without correction, delays in moving toward autonomy and creativity should be expected. Affirmation without correction is a common characteristic of oppressive pedagogy: when a teacher or mentor for example provides shallow feedback like "good job" or "that's nice" without addressing misconceptions or constructively challenging fallacious thinking, mixed messaging catapults the actor into an "entitlement-enablement" loop. Success as a STEM actor is supported most when the path from struggle to progress is made clear with care, collaboration, support, space for failure, correction and endorsements of independence, autonomy, and creative expressions of thought.

This is what we believe leads to transgressive STEM education: that which supports social justice because it attends to the freedom-building needs of teachers and learners. When transgressive practices are normalized, all students have greater access to both windows and mirrors, and a more diverse group of learners develop the knowledge and capacity to tackle the challenges we face now and in the future. STEM equity *is* the fourth dimension—the standard, the benchmark, the goal we must all work to secure for students in the 21st century—it is the least we can do.

Notes

1. Malcolm X delivered these words in 1964 as part of a speech marking the founding of the Organization of Afro-American Unity. Retrieved from http://www.blackpast.org/1964-malcolm-x-s-speech-founding-rally-organization-afro-american-unity
2. Dominant constructs that uphold Western viewpoints include commodification, extractive practices, individualism, competition, Manifest Destiny and the like.
3. Crowley, 2010, p. 1016.
4. Gutiérrez, 2002.
5. The windows-mirrors-doors metaphor appears in many analyses of children's literature; see Botelho and Rudman, 2009. When readers look (or analyze) text as seeing them through

windows, it is an experience of otherness. When analyzed through mirrors, they see themselves. When analyzed through doors, there have opportunities for engagement and interaction—access into and out of the habits and choices of the context being defined.
6. Literally translated as "grain of sand," not only representing our small contribution, but also a documentary film about teachers' grassroots struggle in Mexico. See http://www.corrugate.org/granito-de-arena.html
7. Grounded theory approaches are important for qualitative studies especially when multiple perspectives exist. Additional insight on grounded theory is available through Corbin & Strauss, 1990.
8. The publication crucible is challenging for graduate students. As editors, we consider this opportunity as one of long-distance mentorship where student authors contribute to a discourse outside of their institutions, extending their level of engagement and associations with others like critical friends.
9. Available online at https://www.nap.edu/catalog/13165/a-framework-for-k-12-science-education-practices-crosscutting-concepts#
10. McComas & Nouri, 2016.
11. See STEM Teaching Tools discussion at http://stemteachingtools.org/assets/landscapes/STEM-Teaching-Tool-15-Equity-Overview.pdf
12. The three dimensions of the NGSS are disciplinary core ideas, cross-cutting concepts and science/engineering practices.
13. Examples include Ambitious Science Teaching: https://ambitiousscienceteaching.org/; STEM Teaching Tools: http://stemteachingtools.org/; ESCAPE: http://escape.mspnet.org/; various scholars working on social justice in STEM efforts include Kimberly Tanner of San Francisco State University; Jacqueline Leonard of the University of Wyoming; Nicole Joseph of Vanderbilt University and William Tate of Washington State University among many of the others referenced throughout the text.
14. See "Mobile Fact Sheet" at http://www.pewinternet.org/fact-sheet/mobile/
15. A recent post on Twitter (https://twitter.com/nbrodnax/status/967048356852830208) shared this insight from a conference keynote about technology fairness and accountability.
16. Problematizing discussions of various gender norms and alternative dichotomies of social and physiological characterizations by scientists and science educators is necessary in 21st century identity politics. Basing STEM equity discussions primarily on gender when multiple cultural sub-strata exist within it is counterproductive and easily rejected as useful. Unpacking this idea is beyond the scope of this volume but deserves consideration. For example, the National Science Teacher Association position statement on gender equity (http://www.nsta.org/about/positions/genderequity.aspx) has not been updated since 2003.
17. Today's American Dream Act [H.R. 3839] was proposed in September 2017 as vital legislation in these areas. Notable language shifts to address housing, redlining and food deserts, digital infrastructure and lending practices are described in the revamped legislation, originally introduced in February, 2017 as H.R. 1084.
18. Carello & Butler, 2015.
19. Karplus, 1964.
20. Einstein called for a social justice approach to STEM back in 1945; an audio version of his speech is available online at http://www.openculture.com/2013/06/listen_as_albert_einstein_calls_for_peace_and_social_justice_in_1945.html

21. Banks, 1993.
22. "Righting", the act of challenging views held outside of one's own using disciplined practices of research, is used to describe both formally structured and informal stream-of-consciousness writing that addresses power structures and agency (polity writing). When students are encouraged for example to participate in large authentic discourses that allow for the development of voice and change agency, "righting" of "wrongs" is made more possible. An example of this type of enabling assignment is "Letters to the Next President 2.0" (https://letters2president.org/).
23. Gutiérrez, 2002.
24. *Ibid.*
25. Style, 1988.
26. See Wei-Haas (2016) https://www.smithsonianmag.com/history/forgotten-black-women-mathematicians-who-helped-win-wars-and-send-astronauts-space-180960393/
27. Examples of notable STEMJ figures and events include analysis of the impact of HeLa cells on society, bioethical breaches in sanctioned research like the syphilis studies of Guatemala and Alabama, mineral rights and conquest in Flint and North Dakota or the impact of urban gentrification on public health.
28. At this writing, youth are leading a gun control initiative after a school shooting in Florida that has critical relevance to this discussion. Students are demanding more efficient use of information designed to protect them. They are developing agency and are applying it broadly.
29. As an example, I had students watch *Something the Lord Made*, an HBO film about the life of Vivien Thomas and read excerpts from one of three books Shetterly's *Hidden Figures*, 2016, Skoot's (2010) *Immortal Life of Henrietta Lacks* or Washington's (2007), *Medical Apartheid* in my high school course. Students were able to walk away from that lesson understanding the impact of racial domination in a way that would have been difficult to communicate without having that experience.
30. The first person to discover that insects can hear and alter behavior based on previous experience, and the first African American to receive a graduate degree at the University of Cincinnati.
31. An Omaha doctor widely acknowledged as the first Native American to earn a medical degree
32. An African American cell biologist whose work on melanomas transformed our understanding of cancer
33. An Iranian woman and Stanford mathematics professor, she was the first and to-date only female winner of the Fields Medal of mathematics excellence since it began in 1936.
34. Organic chemist and biomedical engineer of soy-based products, he launched his own pharmaceutical company, was an early pioneer in large-scale synthesis of human hormones from natural products and the second African American inducted into the National Academy of Sciences.
35. Trained by his father to be a carpenter, Thomas worked as a surgical technician at Vanderbilt and Johns Hopkins in the Jim Crow era. Along with his mentor, he developed the tools and procedures used to treat blue baby syndrome in the 1940s.
36. Physicist Leona Woods Marshall Libby was one of the U.S. women who helped to create the atomic bomb in WWII. She was 23 years old at the time.

IMPLICATIONS AND CONCLUSIONS | 247

37. Erin Jackson is the first African American woman to qualify for the long track speed skate events in the Olympics for the team representing the United States. She is a material sciences and engineering graduate from the University of Florida.
38. Shuri is a fictional character in the Marvel Universe. She thrives as a STEM producer in a highly differentiated and culturally rich environment. The 2018 blockbuster film *Black Panther* highlights a laboratory run by the Wakandan princess that is accentuated by virtual and digitally dynamic graphic walls equal in brilliance to the African mural art walls supporting the makerspace. Technology-driven sounds and architecture enhance the open, multipurpose rooms. The mineral rich nation (Wakanda) is protected by five tribes that understand their resources and wealth. Evolving and deviating from traditional belief systems, they protect social, cultural and economic capital in a globally complex political context as a unified and diverse macrosystem. Indigenous knowledges are valued on par with technological advances and modern medicine; both indigenous knowledge and contemporary ideas are integrated into the faith practices and spiritual skeleton that define Wakandan traditions. Although fictional, elements of the characterizations are based on actual African customs, language and beliefs.
39. A significant case in point is provided by consumer/user abuse of artificially intelligent "chat bots" (See https://readwrite.com/2016/03/28/went-wrong-microsofts-tay-ai/ for additional insight).
40. TED history and science standards are available online at https://www.ted.com/about/our-organization/history-of-ted and https://www.ted.com/about/our-organization/our-policies-terms/ted-science-standards, respectively.
41. The National Education Association articulates how social justice is a necessary component of a "diversity toolkit". See http://www.nea.org/tools/30414.htm
42. Corbin & Strauss, 1990.
43. Blaschke, 2012.
44. Tribal sovereignty https://huffman.house.gov/media-center/press-releases/rep-huffman-senate-house-natural-resource-leaders-blast-dakota-access
45. See the "Quiet Crisis Report" online at http://www.usccr.gov/pubs/na0703/na0204.pdf; similar reports on the problems of water pollution in Flint, MI are widely available.
46. See http://www.nytimes.com/2012/07/31/arts/television/disneys-doc-mcstuffins-connects-with-black-viewers.html
47. Teacher-focused mentorship was also evident in this research. The expanded role of teachers as active participants in the research process was part of the research design.
48. STEM student identity is not unique; this model applies generally to a model of "Successful Student" identity where transgressive practices are employed to engage, excite, re-engage and ultimately empower students to become skilled practitioners in a content area.
49. An earlier version of this graphic was presented at the 2015 Hawaii International Conference on Education, based on work done in collaboration with Dr. Timothy Knight (Martin University, Indianapolis, IN) to explain the development of a scholar identity in Black males through mentoring relationships.
50. Ruggs & Hebl, 2012.
51. The "safety net" metaphor is an appropriate reference here; See Duhigg, 2016.

References

Banks, J. A. (1993). Multicultural education: Historical development, dimensions, and practice. *Review of Research in Education, 19*(1), 3–49.

Berry, J. W. (2005). Acculturation: Living successfully in two cultures. *International Journal of Intercultural Relations, 29*(6), 697–712.

Blaschke, L. M. (2012). Heutagogy and lifelong learning: A review of heutagogical practice and self-determined learning. *The International Review of Research in Open and Distributed Learning, 13*(1), 56–71.

Botelho, M. J., & Rudman, M. K. (2009). *Critical multicultural analysis of children's literature: Mirrors, windows, and doors.* New York: Routledge.

Carello, J., & Butler, L. D. (2015). Practicing what we teach: Trauma-informed educational practice. *Journal of Teaching in Social Work, 35*(3), 262–278.

Corbin, J. & Strauss, A. (1990). Grounded theory research: Procedures, canons and evaluative criteria. *Zeitschrift für Soziologie, 19*(6), 418–427. Retrieved from https://www.degruyter.com/downloadpdf/j/zfsoz.1990.19.issue-6/zfsoz-1990-0602/zfsoz-1990-0602.pdf

Corbin, J. M., & Strauss, A. (1990). Grounded theory research: Procedures, canons, and evaluative criteria. *Qualitative sociology, 13*(1), 3–21.

Crowley, R. M. (2010). Transgressive and negotiated white racial knowledge. *International Journal of Qualitative Studies in Education, 29*(8), 1016–1029.

Duffy, D., & Jennings, J. (2010). *Black Comix: African American independent comics art and culture.* New York: Mark Batty Publisher.

Duhigg, C. (2016, February 25). What Google learned from its quest to build the perfect team. Illustrations by J. Graham. *New York Times.* Retrieved from https://www.nytimes.com/2016/02/28/magazine/what-google-learned-from-its-quest-to-build-the-perfect-team.html

Gutiérrez, R. (2002). Enabling the practice of mathematics teachers in context: Toward a new equity research agenda. *Mathematical Thinking and Learning, 4*(2–3), 145–187.

hooks, b. (1994). *Teaching to transgress: Education as the practice of freedom.* New York: Routledge.

Karplus, R. (1964). The science curriculum improvement study. *Journal of Research in Science Teaching, 2*(4), 293–303.

Lee, P. (2016, March 25). *Learning from Tay's Introduction* [Web blog post]. Retrieved from https://blogs.microsoft.com/blog/2016/03/25/learning-tays-introduction/

McComas, W. F., & Nouri, N. (2016). The nature of science and the Next Generation Science Standards: Analysis and critique. *Journal of Science Teacher Education, 27*(5), 555–576.

Ruggs, E., & Hebl, M. (2012). Literature overview: Diversity, inclusion, and cultural awareness for classroom and outreach education. Apply research to practice (ARP) resources. Retrieved from https://www.engr.psu.edu/awe/ARPAbstracts/DiversityInclusion/ARP_Diversity-InclusionCulturalAwareness_Overview.pdf

Style, E. (1988). *Curriculum as window and mirror.* Retrieved from https://nationalseedproject.org/images/documents/Curriculum_As_Window_and_Mirror.pdf

Appendices

Appendix I

Table Appendix I.1: PERC Class Target Behaviors.

The following chart outlines behaviors that occur during different lesson components included in typical PERC classes. An individual lesson might include some or all of these components.

Lesson Component	Student Behaviors	TA Scholar Behaviors	Teacher Behaviors
Do Now	• Enter class on time • Sit in TAS group • Show homework to TAS • Complete Do Now task • Ask TAS and other students questions if needed to complete Do Now	• Enter class in time to set up Do Now materials • Pick up TAS folder with record sheets and materials for the lesson • Greet students as they join group • Record student attendance and homework completion • Encourage students to work on Do Now • Ask scaffolding questions to facilitate Do Now completion	• Greet students and TAS as they enter • Encourage students to join groups and start Do Now • Set up materials for lesson • Speak with students needing individual interventions
Lesson Introduction/ Lecture/ Class Discussion	• Take notes • Ask teacher or TAS questions when confused or to deepen understanding • Answer teacher's questions • Participate in group discussions of questions or problems set by the teacher	• Model note-taking for students • Lead group discussions of questions posed or problems set by the teacher • Respond to student questions and answers with scaffolding questions • If explanation is required, break concepts into manageable chunks • Support ELL students in their home language (bilingual TAS)	• Establish motivation for lesson • Introduce content, concepts, and/or skills of an appropriate quantity and complexity • Pose questions for TAS groups to discuss/problems to explore • Facilitate sharing of group discussions with whole class • Transition effectively between group and whole class work

Lesson Component	Student Behaviors	TA Scholar Behaviors	Teacher Behaviors
Group Work	• Work on task assigned by teacher/TAS • Ask TAS or other students questions when confused or to deepen understanding • Collaborate with other students in completing tasks and developing understanding	• Ask scaffolding questions to facilitate task completion and assess understanding • Respond to student questions and answers with scaffolding questions • If explanation is required, break concepts into manageable chunks • Use own completed work as a reference for supporting students • Encourage students to assist each other productively • Give students appropriate positive feedback • Ask teacher and other TAS for support when needed • Support ELL students in their home language (bilingual TAS)	• Listen to TAS group discussion to assess student understanding and TAS effectiveness • Listen to several exchanges within a group before contributing a question or comment • Model effective questioning and positive feedback • Provide whole class intervention if common misconception is identified across groups • Work with individual students or groups who need additional support
Lesson Closure	• Share out group's ideas/products/ answers from day's lesson • Respond to teacher questions • Ask questions when confused or to deepen understanding • Set a goal • Record homework assignments	• Collect student work • Encourage students to share out and answer teacher's questions • Help students set goals • Ensure that students write down homework • Record student progress	• Ask questions to assess learning from lesson • Prepare students for next lesson • As students to set a goal • Assign homework

Source: Author.

Appendix II: Chemistry Learning for Academic Success in Science—Core/Chemistry Lab and Academic Skills for Success in Science(C.L.A.S.S.) Rubric

Rationale: A clear and simplified tool is needed to support students transitioning from general, college preparatory science classes to accelerated/advanced placement science classes during non-conventional learning.

Goals:
- To provide basis for student evaluation and personal reflection on academic skill development
- To provide qualitative indicators of potential success in year-long course in high level science

Table Appendix II.1: C.L.A.S.S. Indicators *High degree of accuracy (85% of instances)*.

S: Successful U: Unsuccessful

Recommendation Evidence		U/S
Core Chemistry Skills	Use of periodic table or other text resources	
	Quantitative reasoning, measurement	
	Quantitative reasoning, dimensional analysis	
	Nomenclature of variety of molecular and ionic compounds (especially Type II ionic compounds and polyatomic ion nomenclature)	
	Teacher-unassisted stoichiometric calculations	
Lab Skills	Follows lab protocol(s) safely	
	Maintains neat and accurate record of lab activity/lab notebook	
	Demonstrates willingness to clean-up work space	
Academic Skills	Calculator and mental mathematics agility	
	Use of appropriate science vocabulary	
	Works well in a variety of group arrangements	

Recommendation Evidence		U/S
Qualitative *Based on demonstrated social engagement, written & oral communication*	Shows initiative in asking appropriate questions at appropriate times	
	Shows initiative in forming peer groups	
	Seeks help from a variety of resources	
	Uses online learning management system appropriately	
Quantitative *Based on formal assessments*	Performs at a high level on standardized assessments	
	Completes homework and classwork assignments with a high degree of accuracy	

Additional Recommendation Evidence/Comments:
Source: Author.

Appendix III: COMPASS Framework of Equitable Teaching

The COMPASS Framework for adult stakeholder evaluation was developed by Joy Barnes-Johnson for NJ GEAR UP, a formal out-of-school time academic support program, the COMPASS framework was developed as part of an initiative to prepare teachers and counselors/student advisors in mostly urban high-needs schools for educational reforms. Over the course of five years, the framework was used to guide professional development of educators and included a research-based observation protocol for program evaluation.

Table Appendix III.1: COMPASS Framework for Adult Stakeholder Evaluation.

	Indicator Description Evidence—Teachers	Indicator Description Evidence—Counselors
C	Community relevant activities parallel important topics within the student community	Community relevant projects that are aligned with student goals, student individual communities, and student led formation of *identity development*.
O	Organized learning flows well and follows an appropriate plan that is available for review	Counselors design learning activities by employing a hierarchy of the American School Counselor Association standards, student personal education plan and student's personal co-developed goals, which can be derived from the standards.
M	Meaningful tasks are academically important and connect students with content	Tasks are project-based and conceived with counselor's support and the projects relate to students' career/passion interests.
P	Previewed activities are advance selected materials that demonstrate instructors' prior knowledge	Learning resources are developed to inform students' projects and instructors are versed on information prior to leading workshops.
A	Academic habit-forming activities foster research, academic independence	Student becomes co-facilitators. Students follow an *inquiry approach* and employs reflection, discussion, and evidence-based reading/writing during workshops.

	Indicator Description Evidence—Teachers	Indicator Description Evidence—Counselors
S	Skill building activities clearly connect to appropriate standards	Skills connect with goals identified by the American School Counselor Association standards for *personal development*.
S	Student interest motivated activities engage students as evidenced by students on task and show genuine interest in learning; Evidence of differentiated instruction (skill, interest)	Students modify counselors' "big idea" goals based on their individual projects.

Source: Author.

Appendix IV: STEM Engagement Rubric

In the mixed-ability classroom, a variety of strategies are used to accommodate students as they learn. When students are able to demonstrate pro-social behaviors it is easier to gauge whether these behaviors will lead to positive teacher-independent academic outcomes. Literacy strategies that involve reading, writing, listening, speaking, observing and drawing (including doodling) could all be used by the student to demonstrate understanding. Building from a writing technique called CAP, students could use literacy strategies to provide evidence of STEM learning. They were invited to score themselves with the rubric and submit it as evidence that they were in-fact engaged, even if on the surface, it may not have seemed apparent. Used for formal and informal class activities, the only time when students "earned" a zero was when they made no attempt to engage class discussion or activities. Using this rubric as a guide, even if students posted/shared on social media (in real time or afterward), as long as the time/date-stamped "text" was relevant to the discussions and activities of the day, they could use the information as verification of learning.

Table Appendix IV.1: STEM Engagement Rubric.

Score	Description	
0	No attempt: student does not answer question; no response recorded	*Unengaged*
1	Off task: student does not address the questions; student repeats question without responding to task; student includes irrelevant information in response; "stream of consciousness" writing; lacks cohesion or singularity of thought	*Entering*
2	Some important points included: student may recite question but extends response to include additional details; student uses line numbers or other references but does not attend to question beyond statement of these references; student uses limited content-specific vocabulary to convey ideas; obtrusive errors disrupt flow of text; general sense of a lack of revision	*Emerging*
3	Most important points included: student uses line numbers and integrates references and/or relevant external information to strengthen written response; student synthesizes details provided in text to build an appropriate argument/make claims in response to the question; student uses a range of content-specific and general vocabulary to convey ideas; clear evidence of revision and clarification beyond "stream of consciousness" writing is apparent; if using variation techniques, student uses parallel literary constructs and conventions with direct attribution/citation of original text	*Engaging*

Source: Author.

Appendix V: STEM-21 Curriculum

Portfolio Topics

Quarter 1: Design for Disaster

Quarter 2: Forces, Motion, Energy & Structure

Quarter 3: Human Impact

Quarter 4: Alternative Energy

Unit	Topics	Enduring Understanding	Skills
Defining STEM in the 21st century Essential Question 1: Why are science and technology the most appropriate design tools for preparing to deal with disaster? Essential Question 2. How are modern technologies transforming the distribution and quality of information generated about science, engineering and mathematics? Essential Question 3. Why is it necessary to address societal/sociological questions/problems through research in science and technology?	• Sociology of STEM (Progress) • Research methods (resource quality) • Scientific and engineering processes • Quantifying the material world • Characteristics of digital information/technology • Cost/benefit analysis • Careers in emergency manage-ment • Scientific methodology • Macroscopic organization of matter • Food safety • Physical safety • Information security • Properties of material • Energy types	Students will contemplate each of the following statements: • Technology is used to help humans explain, reproduce and predict various phenomena. • Science uses data and information collected over time to explain, reproduce and predict various phenomena. • Models help explain and predict various phenomena. • Mathematical relationships and conventions are used to explain and predict various concepts. • The behavior of objects in the physical world are determined in part by the materials used to construct/define those objects. • Sociological, technological and physical structures exist in the 21st century that distinguish it from other periods of time and advancement.	Students can: • Define characteristics of human activity and enterprise that create a need for scientific and engineering endeavors • Measure various objects using multiple instruments, unit systems and techniques • Analyze and interpret informational text (technical and non-technical) • Describe sociological phenomena and structures observed over the last 100 years that have influenced 21st century understandings

(Continued)

Table Appendix V: (*Continued*)

Unit	Topics	Enduring Understanding	Skills
Matter: Structure & Function Essential Question 1: How can one explain the structure, properties and interactions of matter? Essential Question 2. How do substances combine or change (react) to make new substances? Essential Question 3. How does one characterize and explain these reactions and make predictions about them?	• States of matter • Bulk vs. microscopic vs. nanoscale • Wave-particle duality of nature • Material Science Hierarchy (Element > Compound > Material > Product) • Periodic table (Metal/non-metal; stable/radioactive nuclei)	Students will contemplate each of the following statements: • Matter is studied and understood at various organizational levels (macroscopic, microscopic and nanoscale). • The particle nature of matter (structure) helps determine its function and behavior. • Matter changes as a result of energy.	Students can: • Draw models to show the organization of matter across various scales • Describe characteristics of each state of matter • Explain how energy transfer (input/output) impacts the behavior of matter • Use the periodic table to predict behaviors of certain types of material
Forces & Motion Essential Question 1. How do we know that forces behave in predictable ways? Essential Question 2. How is constant motion distinguished from static and changing motion states?	• Constant speed • Average speed • Equal and opposite [balanced] forces (1st law) • Constant force (2nd law) • Applied forces (including pressure)	Students will contemplate each of the following statements: • Materials used to create objects influence the motion (movement) of those objects because of the particle nature of matter. • Newton's laws help explain the macroscopic motion of most objects. • Conservation laws of energy and momentum help explain and predict movement.	Students can: • Graphically represent forces as vector quantities • Graphically represent mechanical motion • Measure distance • Convert distance units • Construct a Newtonian vehicle

Unit	Topics	Enduring Understanding	Skills
Essential Question 3. Why is the relationship between mechanical systems and other systems (biological, chemical) important for understanding structure, function and change?		• Changes (transformation) in energy accompany changes in matter.	
Research & Development (CEROw or ZERO) Essential Question 1: How is the BEST design response to a real-world challenge created by disaster identified? Essential Question 2. Who is on the response team? Why? Essential Question 3. What can citizens use from their everyday life to create a solution? Essential Question 4. What should citizens know about the cost/risk and benefit of this solution?	• Research methodologies involving databases • Research source quality • Engineering design tools	Students will contemplate each of the following statements: • Global challenges/real world problems can be verified as important when confirmed from multiple perspectives. • Contributions to solutions for global challenges are made from many different stakeholders who each must consider costs, risks and benefits. • Models developed as a result of research processes allow scientists and engineers to make claims based on evidence that potentially lead to solutions to challenges/problems.	Students can: • Use various instruments to make measurements accurately and with precision (reliability) • Conduct independent research about a scientific topic using primary source technical materials • Develop a digital or physical model of a real-world problem of global significance • Write a technical report of findings based on an independent investigation of a real-world problem

Summative Assessment: STEAM Rising—Unsung Heroes of STEM Community Event

Students study the lives and sociological contexts that have produced some of the greatest multicultural thinkers over the last 200 years. Students share character sketches and essays written about men and women from around the globe, many of whom remain "hidden figures" in the academic and artistic landscape of people who are more well known. A multi-level report of the lives and experiences of these STEM actors is treated as a summative evaluation of student understanding about the diverse contributions (diversity of contributors) that define human enterprise, innovation and discovery.

Components

A. STEAM report
   1. Science advancement profile
   2. Technology advancement profile
   3. Engineering advancement profile
   4. Art profile

B. Sociological system report
   1. Population profile
   2. Data representation of sociological system

C. Family/Community Oral Presentation
D. Wiki/Biographical Essay (Example https://stemunsung.wikispaces.com/)
E. Timeline with at least five world events that frame narrative

Index

21st century, 4, 6, 14, 19, 25, 27, 38, 50, 61, 103, 112, 172, 174–175, 178–179, 181–182, 225–226, 231, 234–235, 244–245, 257
4th dimension, 239–240

A

AAAS, 5, 33, 137, 206
Abbott, 27–28, 47
Accountability, 24–26, 34–35, 45, 65, 101, 112, 173, 219, 245
Achievement gap, 9, 14, 16, 27, 60–61, 75, 202, 207, 223–224, 227
Action research, 5, 17, 37, 46, 97, 176, 184
Agency, 4, 8, 19, 22–23, 25, 33, 50, 56, 71, 79, 96–97, 116, 139, 176, 231–235, 237–240, 242, 246
America COMPETES Act, 25
ASCD, 15, 29, 162
Aspirations Framework, 8, 96, 116

Authentic, 3, 25, 50, 59, 63, 70, 97, 110, 115–116, 130–132, 166, 175, 180, 192, 196, 207, 231, 241–242, 244, 246

B

Backward design, 168
Benchmark, 244
Bias, 2, 9, 172, 226
Biology, 6, 35, 44, 80–81, 98, 103, 127, 184, 194, 205, 214, 217
Black Panther, 234, 247
Border crossing, 4, 57–58
Boundaries, 4, 20, 39, 56–57, 141, 185

C

Capacity building, 35, 48, 110, 164, 205, 207, 210, 212, 214, 217–219, 221, 240
Cities, 29, 140, 190, 206

Civil rights, 17, 29, 92, 229
Classroom management, 66, 105
Climate change, 24
Co-Teaching, 64–65, 74
Coding, 18, 44, 46, 78, 212, 237
Collaboration, 4, 14, 31, 36, 39, 51, 96, 100, 108, 144, 168, 197, 206, 209, 217, 238, 244, 247
College readiness, 81, 93, 96, 98, 111, 193, 203
Colorado, 79, 138–139, 141, 144, 146, 161, 167, 206, 208–211, 217, 220, 222–223
Communication, 14, 24–25, 36, 43, 49, 72, 123, 127, 144, 163, 173, 175–177, 179, 189, 193, 195–197, 199, 202–203, 230, 232, 243, 253
Community college, 74, 76, 164, 206, 208, 216, 218, 223
Community relevant, 32, 45, 58, 254
Community-based, 3, 58, 87, 137, 164, 206, 218
Constructivism, 31, 39, 50
Cooperative learning, 100, 167
Critical, 4–5, 8–9, 11, 14, 16, 23–24, 26–27, 29–32, 39, 42, 45, 47–48, 51–53, 81, 91, 100–102, 105, 113, 119, 121, 131, 144, 166, 173, 181, 183, 186, 191, 194, 196–197, 200, 209–210, 218, 226, 230, 232, 234, 245–246, 248
Cross-cutting concepts, 171, 245
Crowdsourcing, 23, 51–52
Cultural competency, 189, 193
Culturally relevant, 30, 32, 45, 63, 65, 71, 73, 121, 195, 204, 230, 236
Culturally responsive, 9, 14–15, 29, 32, 45, 52, 54, 186
Curriculum, 6, 13, 24, 27–29, 32, 34, 51, 53–54, 61, 70–71, 77–78, 80–82, 110, 120, 128–129, 139, 141–143, 145–147, 150, 152, 154–155, 157–159, 165–166, 168, 170–172, 174, 176, 183, 186–187, 193, 210, 232, 243, 248, 257

D

Data analysis, 118, 126, 130, 164, 172, 212, 223
Deficit, 3, 130, 167, 219, 226–227
Differentiation, 36, 166, 171, 182, 185
Digital divide, 25
Disability, 27, 46, 184
Disciplinary core ideas, 3, 12, 245
Diversity texts, 37, 45, 49, 177
Drone, 145–146, 155, 161

E

Ecology, 91, 117, 124, 131, 133
Efficacy, 24, 34, 37, 41, 49, 51, 81–83, 91–92, 116, 157, 169, 173, 215, 223–224, 231, 238, 243
Elementary, 7, 24, 31, 33–34, 47–49, 51, 53–54, 59, 84, 92, 141, 167, 205, 224
Engineering, 1–2, 12, 15–16, 24, 40–42, 47, 51–53, 62–63, 68, 75, 84, 86–87, 92, 94, 115, 123, 126–127, 130, 136, 138–149, 151–162, 171–175, 178–180, 183, 197, 202–203, 207, 210, 214–215, 223, 231–232, 235, 245, 247, 257, 260
Environmental racism, 230
Equitable science teaching, 9, 20, 28, 31–32, 35, 37, 40, 45, 51–52, 168, 186, 226, 230
Equity Metric, 8–10, 140
ESEA, 24–26, 46
ESSA, 24, 26, 42, 47, 191
Essential questions, 13, 15, 168, 173
Essential skills, 209
Evaluation, 5, 10, 42, 46, 52, 72, 92, 96–97, 137, 144, 159–160, 166–167, 173, 175, 183, 185, 187–188, 192, 199, 211, 228, 252, 254, 260
Excel, 62, 83, 126
Experiential, 20, 39, 45, 130
Experiment, 78, 165–166, 170, 233

F

Food, 24, 38, 78–80, 85, 87, 144, 178, 233, 236–237, 245
Formal education, 11, 13, 220–221
Formative assessment, 100, 105, 111, 113
Fourth dimension, 5, 229, 235–236, 244
Framework, 5–6, 8, 14, 16, 18, 25, 28, 33–35, 41–42, 45, 49, 55, 63, 74, 77, 81, 92, 96, 100, 116, 139, 141, 144–145, 164, 176–177, 184–187, 205, 209, 218, 224, 229–232, 235, 237, 239–240, 245, 254
Funds of knowledge, 15–16, 37, 45, 221

G

Grounded theory, 223, 228, 245, 248

H

HBCU, 7, 56, 72
Health, 10, 12, 24, 34, 81, 83, 112, 163, 182, 188–203, 231, 236–240, 246
Heterogeneous, 168
Heutagogy, 173, 186, 237, 248
High school, 6–7, 14, 18, 20–21, 26, 33, 46–47, 53, 56, 61, 90, 94, 96, 98–99, 109, 111, 113, 115, 117, 119, 121–125, 127, 129, 132, 135, 139, 141, 143–144, 161–163, 167, 169, 188–189, 192–198, 202–203, 214, 216, 218, 246
High stakes, 35, 96, 98–99, 109
High-need, 160
Hip-Hop, 17–18, 23, 37–38, 43, 46, 55, 57, 59–75, 167, 230, 232, 239–240
Hispanic Serving Institution (HSI), 208
hooks, b., 15, 248

I

Identity, 10, 21, 34, 39, 49–50, 58–59, 66, 71, 87, 116, 135–136, 162–163, 165, 169, 176, 194–195, 197, 218, 223, 238–243, 245, 247, 254
Imminent domain, 21
In-service, 3, 31, 53, 97, 168, 170, 184
Inclusion, 3, 5–6, 32, 36, 41, 116, 167, 193, 207, 218, 224, 228, 248
Income, 38, 77–78, 80, 87, 92, 94, 117, 119, 139–143, 145, 158, 191, 195–196, 208, 223
Index, 8, 13, 16, 48, 52, 161, 185–186, 191, 203, 222
Indigenous, 9, 13, 32, 50, 116, 190, 234, 236, 247
Industry, 1–2, 11, 33, 179, 181, 206, 230, 236
Informal education, 128
Inquiry, 3, 15, 20, 33, 38, 52, 80, 82–84, 113, 173, 175, 177, 186, 195, 200, 210, 223, 238, 240, 254
Intersectionality, 39
Invisibility, 6, 16, 32, 54
ISTE, 161
Iteration, 157, 206

K

K-12 framework, 144
Kinesthetic learning, 67
Knowledge of self, 63–64, 69–70

L

Latinx, 13, 58, 60, 62, 79, 83, 89–90, 96, 99, 101–102, 114, 121–122, 145, 191, 196, 209, 219, 234
Leadership, 4, 8, 13, 15, 47, 50, 52, 85–86, 90, 93, 97, 100, 108, 110, 112, 114,

117, 119–120, 122, 125, 127–128, 130, 132–133, 136, 140, 145, 162, 169, 173, 196, 199–200, 204, 209, 233
Learning environment, 35, 56, 195–196, 242
Legacy, 20, 23, 29, 32, 48, 189, 208, 221, 231, 236
Legislation, 24, 26–27, 47, 245
Lesson study, 163, 165, 170, 183–184, 186
Liberatory, 237
Licensure, 164, 167, 206, 210, 213, 215–217
Life science, 163, 194
Linear, 184, 242
Linguistic and cultural diversity, 4
Literacy, 16, 22, 33, 41, 49, 74, 92, 112, 115–116, 144, 147, 150, 160, 163, 165, 167, 169, 171, 173, 175, 177–179, 181, 183, 185, 187–189, 191–192, 194, 196–197, 199, 202–203, 205, 209–210, 224, 231–232, 239–240, 256
Lyndon B. Johnson, 24

M

Mainstream, 5, 34, 48
Makerspace, 39, 45, 51, 247
Marginalized, 7, 17, 31, 41–42, 48, 55, 60–61, 63, 68, 70, 72, 97, 190–191, 200, 221, 231, 233
Mathematics, 1–2, 10, 12, 14–16, 24–26, 34–35, 40–42, 44, 48–50, 52–54, 62, 68, 72, 90, 98–100, 109, 144–147, 161, 171–175, 178–180, 183, 186, 199, 202, 205, 213–215, 219, 222, 224, 228, 235, 246, 248, 252, 257
Media, 3, 23, 39, 44–45, 48, 166–167, 184–185, 202, 206, 220–222, 228, 233–234, 236, 247, 256
Medical education, 194, 196, 199, 201
Mentor, 56, 115, 123–124, 131–132, 197, 240, 243–244, 246
Meritocracy, 26, 172, 235
Microclimates, 197, 231
Mindfulness, 81

Motivation, 94–95, 97, 105, 138, 141, 143, 152–153, 156, 158–163, 169, 196, 203, 223, 239–241
Multicultural education, 17, 25, 28–31, 38, 41, 51–53, 74, 230–231, 233, 236, 248
Music, 37–38, 43, 50, 63, 66, 68–70, 74, 135

N

Narrative, 12, 14, 26, 34, 41, 162, 212, 231, 260
NARST, 23, 49
National Center for Education Statistics, 53, 110, 112, 140, 160–161
National Oceanic and Atmospheric Administration (NOAA), 118
National Science Foundation (NSF), 119, 141, 209
NCLB, 24, 48, 53
NCTM, 224
Near peer, 110, 140
Network, 63, 111, 115, 119, 127, 145, 211, 216, 219, 223, 226, 237, 241–242
Neuroscience, 127, 236
Neutrality, 24, 32
Next Generation Science Standards (NGSS), 144, 171, 191, 226
Non-dominant, 3, 9–10, 17, 22, 24, 34, 39, 48, 60–62, 77, 89–90, 93, 99, 163, 200, 227, 234, 237
Notebooks, 142, 150, 158
NRC, 171, 174
NSF, 16, 26, 43, 75, 91–92, 95, 98, 114, 119, 122, 129, 141, 203, 205, 209–210, 218, 222, 224
NSTA, 23, 49–51, 111–112, 162, 176, 185, 245

O

Opportunity gap, 14, 80, 101, 208, 224, 227
Opportunity to Learn, 6, 15–16, 32, 52, 65, 79, 97, 140

Out of School, 9, 78–80, 83, 89–90, 141, 204
Outreach, 114–115, 123, 139, 153, 199, 206, 216, 223, 248

P

Partnership, 14, 98, 109, 118, 128, 139, 188, 199–200, 209
Pathways, 2, 8, 57, 79, 140, 160, 167, 170, 175, 188–189, 191, 193, 195, 197, 199–201, 203, 208–209, 212, 214, 216–218, 223, 232, 237–238, 242–243
PBL, 202
Pedagogy, 4–5, 9, 11–12, 15, 17–18, 30–32, 34, 41, 45, 48, 52–53, 55–57, 59, 61–71, 73–75, 100, 167, 173, 207, 221, 226, 231, 234, 237, 242, 244
Performance Expectations, 229
Pipeline, 43, 93, 98, 109, 137, 163, 188–189, 192–200, 207, 213, 216, 242–243
Place-based, 4, 116, 130–131, 133, 139, 192, 194, 238
Portfolio, 173, 175–176, 183, 257
Positionality, 5, 16, 236
Post-secondary education, 145, 163, 189, 192, 194, 198–199
Poverty, 9–10, 24–26, 35, 49, 52, 74–76, 190, 192, 195, 204, 207–208, 234
Pre-service, 3, 5, 51, 53, 217
President Obama, 43, 62, 75
Princeton (NJ), 1, 19, 165, 225
Privilege, 22, 42, 97, 232
Professional development, 48, 70, 83–84, 92–93, 102–103, 106, 108–109, 119, 128, 145, 167, 169–170, 190, 218, 228, 254
Professional learning community, 97, 169
Professional organizations, 3
Public schools, 1, 19, 112, 117, 121, 128, 140, 165, 167, 169, 194, 202, 225
PWI, 56, 72

Q

Qualitative, 16, 97, 101–102, 111, 139, 153, 156, 175, 192, 211–212, 223–224, 237, 245, 248, 252–253
Quantitative, 130, 132, 139, 153, 156, 175, 179, 192, 211, 252–253

R

Racial diversity, 121, 137, 196
Racism, 9, 116, 172, 207, 230, 234
Reality, 63, 71, 74, 167, 191, 197, 227
Reflexive, 5, 35, 190, 223, 229
Reform, 4, 8–11, 15–16, 29–31, 33, 46, 52, 163, 166, 169–170, 174, 186–187, 199, 202
Relationship, 13, 26, 45, 50, 56, 79, 85–86, 90, 104, 106–108, 169, 183, 237, 243, 259
Religion, 49
Religious, 44, 52
Restorative, 25
Risk-taking, 169
Role model, 104–105
Rubric, 33, 96–97, 175–176, 252, 256

S

Scaffolding, 20, 95–96, 103, 124, 250–251
Scholarship, 7–8, 23, 28–30, 39, 52, 164, 205, 210, 216, 218
School engagement, 194–195
Science identity, 71, 136, 194
Science knowledge, 192, 194
Secondary, 1, 20, 24, 27, 47, 56, 100, 111, 113, 115–117, 145, 163–165, 167, 172, 186–187, 189, 191–192, 194, 198–199, 205–206, 215–216, 218, 225, 238, 240
Senate, 247
Sense-making, 24, 238
SES, 34
Social congruence, 196, 200, 204

Social justice, 3, 18, 32, 34, 42, 47, 79–82, 86, 89–92, 190, 221, 226, 230–231, 233–234, 236–237, 244–245, 247
Social media, 23, 39, 45, 184–185, 228, 236, 256
Socioeconomic status, 232
Special needs, 34, 83
Spiritual, 32, 227, 232, 234, 247
Stakeholder, 4, 20, 25, 164, 174, 206, 211, 254
Standards, 3–6, 8–13, 15–18, 28, 31–34, 36, 38, 44, 47–49, 54, 72, 74, 83–84, 101, 108, 110–111, 113, 144, 159, 168–169, 171, 175–176, 185, 191, 193, 226, 228–229, 231, 235–239, 247–248, 254–255
STEBI, 82
Stereotype threat, 208
Student Learning Outcomes, 1, 143, 173–174, 221
Student-centered, 46, 96, 98, 109, 130
Subtexts, 237–238
Sustainable, 4, 25, 37, 98, 109, 115, 170
Systemic, 9–10, 15, 52, 54, 58, 61, 116, 136, 200, 207, 209

T

T-STEM, 82
Teacher preparation, 32, 58, 121, 167, 210, 218
Technology, 1–3, 10–12, 16, 23–24, 26, 32, 34, 40–43, 46–47, 49, 52–53, 62–63, 68, 75, 86–87, 92, 113–114, 139, 141, 144, 147, 157, 160–161, 165, 168, 171–175, 177–179, 183, 185–187, 202–203, 206–207, 210, 230, 235–236, 245, 247, 257, 260
The United States Department of Education, 62
Three-dimensional learning, 3, 38
Tracking, 40, 43, 119, 122, 193
Traditional, 8, 22–23, 34, 42, 44–45, 62, 65, 69–71, 96, 99–100, 103–104, 106, 109, 135, 167, 170, 175, 196, 209, 217, 221, 229, 232–233, 238, 240, 242, 247

Transdisciplinary, 9, 115, 228
Transgressive, 4–6, 10–12, 23–24, 39–42, 45, 79, 97, 163, 165, 167–169, 172–174, 176–178, 185, 199, 221, 226–227, 229, 231–234, 237–238, 240, 243–244, 247–248
Transgressive practices, 6, 11, 163, 185, 226, 231, 233, 237–238, 240, 244, 247
Transgressive teaching, 4, 10, 79, 97, 167, 199
Typology, 29, 31

U

UAV, 139, 142–152, 155–157, 161
Underachievement, 9, 15, 162
Undergraduate, 5, 120–123, 130, 132, 163, 194, 208–209, 219
Unmanned Aerial Vehicles, 139, 142
Urban, 6, 14–18, 27–29, 31–32, 34–35, 37, 39, 51–55, 59–63, 66–68, 70, 73–76, 78–80, 91–92, 95–99, 101, 103, 105, 107, 109, 111, 113, 118–119, 134, 164, 192, 195, 202–206, 208–210, 224, 230, 236, 246, 254
Utility value, 142–143, 146, 150, 153, 155–156, 161

W

War on poverty, 24–26
Western science, 38
White House, 43, 62, 75, 110, 113
Whiteness, 226
Windows and mirrors, 244
Wrinkle in Time, 234

Y

YouTube, 23, 46, 69, 185

Conceptual Framework Theorists Index

[Frequently referenced conceptual frameworks by theorist (theory), alphabetical listing]

A

Atwater, Mary (*Critical race theory of science education*), 31, 48, 50, 53

B

Banks, James (*Multicultural education*), x, 29–31, 41, 47, 51, 231, 246, 248

D

Deloria, Vine (*Global social theory; Indigenous community*), 13–14, 136–137
Delpit, Lisa (*Culture of power*), 116, 222

E

Emdin, Chris (*Hip-Hop pedagogy*), 65, 72–74

G

Gay, Geneva, (*Culturally responsive teaching/ pedagogy; empowerment pedagogy*) 29, 47, 52
Gorski, Paul (*Counter deficit theory; multicultural education*), 201–202, 222
Gutiérrez, Rochelle (*Critical mathematics; dimensions of equity/equity framework; dominant mathematics; equitable mathematics education, gap gazing*), 186, 226, 231, 244, 246, 248

H

hooks, bell, (*Transgressive teaching*), 4, 10, 13, 15, 185, 232, 237, 242, 248

K

Kahle, Jane Butler (*Equity metric*), 8–10, 14–15, 48–49, 52–53, 74

Karplus, Robert (*Educative science play*), 231, 245, 248

L

Lee, Ohkee (*Equitable science education; linguistic diversity*), 48, 51, 53, 177

Luykx, Aurolyn (*Cultural congruence science and language*), 48, 53

M

McLaren, Peter, (*Multicultural education*) 4, 15

Moll, Luis (*Funds of knowledge*), 13, 15–16, 37, 221

R

Rogoff, Barbara (*Sociocultural theory*), 13, 15

S

Sleeter, Christine, (*Multicultural education*) 29, 222

Style, Emily Jane, (*Windows-Mirrors*) 232, 246, 248

sj Miller & Leslie David Burns
GENERAL EDITORS

Social Justice Across Contexts in Education addresses how teaching for social justice, broadly defined, mediates and disrupts systemic and structural inequities across early childhood, K–12 and postsecondary disciplinary, interdisciplinary and/or transdisciplinary educational contexts. This series includes books exploring how theory informs sustainable pedagogies for social justice curriculum and instruction, and how research, methodology, and assessment can inform equitable and responsive teaching. The series constructs, advances, and supports socially just policies and practices for all individuals and groups across the spectrum of our society's education system.

Books in this series provide sustainable models for generating theories, research, practices, and tools for social justice across contexts as a means to leverage the psychological, emotional, and cognitive growth for learners and professionals. They position social justice as a fundamental aspect of schooling, and prepare readers to advocate for and prevent social justice from becoming marginalized by reform movements in favor of the corporatization and deprofessionalization of education. The over-arching aim is to establish a true field of social justice education that offers theory, knowledge, and resources for those who seek to help all learners succeed. It speaks for, about, and to classroom teachers, administrators, teacher educators, education researchers, students, and other key constituents who are committed to transforming the landscape of schools and communities.

Send proposals and manuscripts to the general editors at:

sj Miller	sj.Miller@colorado.edu
Leslie David Burns	L.Burns@uky.edu

To order other books in this series, please contact our Customer Service Department at:

(800) 770-LANG (within the U.S.)
(212) 647-7706 (outside the U.S.)
(212) 647-7707 FAX

or browse online by series at:

WWW.PETERLANG.COM